Ernane Cunha de Lima

Increased excavator tip life

Ernane Cunha de Lima

Increased excavator tip life

Increasing the productivity of mining equipment

ScienciaScripts

Imprint
Any brand names and product names mentioned in this book are subject to trademark, brand or patent protection and are trademarks or registered trademarks of their respective holders. The use of brand names, product names, common names, trade names, product descriptions etc. even without a particular marking in this work is in no way to be construed to mean that such names may be regarded as unrestricted in respect of trademark and brand protection legislation and could thus be used by anyone.

Cover image: www.ingimage.com

This book is a translation from the original published under ISBN 978-3-330-99633-5.

Publisher:
Sciencia Scripts
is a trademark of
Dodo Books Indian Ocean Ltd. and OmniScriptum S.R.L publishing group

120 High Road, East Finchley, London, N2 9ED, United Kingdom
Str. Armeneasca 28/1, office 1, Chisinau MD-2012, Republic of Moldova, Europe
Managing Directors: Ieva Konstantinova, Victoria Ursu
info@omniscriptum.com

Printed at: see last page
ISBN: 978-620-8-59753-5

To Cleto Lima and Neusa Cunha, my parents, and Nayara, my daughter.

ACKNOWLEDGEMENTS

To Professor Maria Celeste Monteiro de Souza, for her encouragement and support in the preparation of the work related to my Master's degree.

To the Federal Technological Education Centre of Minas Gerais - CEFET, in the person of Ivete Peixoto Pinheiro Silva, Ezequiel de Souza Costa Júnior, Ângela de Mello F. Guimarães and Elaine Carballo S. Corrêa, for their support in the preparation of my work.

To Neusa Cunha de Lima, my mother, Cleto Alves de Lima, my father, Evandro Cunha de Lima and Cleto Alves de Lima Júnior, my siblings, Nayara Oliveira Lima, my daughter, and my girlfriend Vânia Guedes for their understanding during the writing of this dissertation.

To Dr. Denilson José do Carmo (Researcher and Instructor at SENAI-CETEF) for his commitment to finding solutions to our problem and to the engineer Elifas Levi for his support in the work carried out.

Whoever you are, whatever social position you have in life, high or low, always aim for a lot of strength, a lot of determination and always do everything with a lot of love and a lot of faith in God, that one day you'll get there. Somehow you'll get there.

Ayrton Senna

SUMMARY

In industry in general, and especially in the mining industry, one of the biggest concerns is related to the loss of metal (mass) caused by abrasive wear mechanisms and impacts that cause plastic deformation in parts and equipment, as these represent one of the main factors in capital depreciation, loss of productivity and sources of maintenance costs. In addition to influencing direct production costs due to the need to replace or recover worn parts, it also influences indirect production costs due to the need to oversize components and production limitations due to deteriorated equipment, as well as often unforeseen interruptions. In view of the above, this work aims to develop and comparatively analyse the wear resistance of large excavator tips with and without hard material liners in different mining soils. The methodology used begins with choosing the equipment with the highest cost in terms of changing the bucket tip. Once this piece of equipment has been chosen, tests begin with the original tips that came with the purchase of the equipment, going through changes in the shape of the soil penetration, changes in the mechanical and chemical characteristics of the tips and, in parallel, the final test with hard material applied superficially through the welding process. With the results obtained, it was possible to see that the tip developed by a partner supplier proved to be an interesting alternative, since its wear resistance was 100 % higher than the original tip supplied by the equipment manufacturer. This was not enough to achieve the goal of the tips having a useful life of 250 hours, but the useful life was doubled compared to the initial useful life. This resulted in a gain in reals per hour worked, reducing the cost/hour worked indicator by around 66 per cent. There are other intangible gains, such as a reduction in machine downtime for changing tips, a reduction in ergonomically poor maintenance services and an increase in equipment productivity, as the tip remains longer and wider for longer, increasing penetration and strength for extracting rock from slopes or soils. The tests were carried out on the entire mine, which is made up of various types of soil, each with its own characteristic abrasiveness/impact and grain size, ranging from loose material (friable) to rocks of great volume and weight (lamellar). This huge variation in rock materials provides a range of service lives from 50 to 300 hours.

Keyword: Mining. Excavator tip. Abrasion wear. Impact wear.

Summary

CHAPTER 1	**6**
CHAPTER 2	**8**
CHAPTER 3	**10**
CHAPTER 4	**11**
CHAPTER 5	**53**
CHAPTER 6	**65**
CHAPTER 7	**82**
CHAPTER 8	**83**

CHAPTER 1

1.1 - INTRODUCTION

In many industrial sectors, abrasive wear is the villain of production, i.e. it is responsible for surface damage to equipment. This damage causes a considerable reduction in productivity due to constant downtime to replace worn parts and components.

In the mining sector, there is often a need to manufacture machine parts or components that must have good resistance to abrasion and wear, combined with good impact resistance. In this case, certain special properties need to be imparted to the outer layers of these components in order to considerably improve their resistance to fatigue and abrasion.

One of the possible ways to combat abrasive wear on components used in mining is to use the method of depositing a special wear-resistant alloy, called a hard coating, on the surface subject to deterioration. Welding processes are commonly used to deposit this hard coating.

Various welding techniques, such as the oxy-acetylene gas welding process (OAW), the gas metal arc welding process (GMAW), the coated electrode electric arc process (SMAW), and the submerged arc process (SAW) can be used for hardfacing any part. The most important differences between these techniques lie in the efficiency of the weld, the dilution in the weld plate and the cost of manufacturing the consumable weld. The coated electrode process (SMAW) is used due to its low cost and easier application, while flux-cored wires have higher productivity and better weld quality (Buchely et al, 2005).

The research aims to study a commercial electrode in terms of its chemical composition, microstructure, hardness and resistance to abrasive wear.

The general aim of this work is to develop and comparatively analyse the wear resistance of large excavator tips that are responsible for excavating material on mine benches, with and without hard material linings. It is hoped that, with the results obtained, it will be possible to provide better information on which type of tip has the longest useful life and the best cost-benefit ratio.

The justification and relevance of this work are presented in Chapter 2, pointing out the need for development.

This work contains a general objective and three specific objectives, which are described in Chapter 3.

A literature review is described in Chapter 4, where some technical concepts are presented.

A detailed description of the experimental procedures and the equipment used for the study is given in Chapter 5.

The results of the work and the discussions about it are presented in Chapter 6.

Finally, chapter 7 presents the conclusions.

CHAPTER 2

2.1- JUSTIFICATION AND RELEVANCE

Excavation, loading and transport operations in a mining company are the most critical and complex operations within the mining processes, since they represent approximately 40 to 60 per cent of the operating costs of all related processes, according to Rodrigues (2006), apud Quevedo (2009).

Therefore, this study is justified by the relevance that the financial savings of replacing tips, components and equipment in a mining industry can provide, as well as having an impact on various sectors of the company such as sales, maintenance, quality control, accounting, finance and others.

The high cost of mining parts is mainly due to wear and tear caused by the varying abrasiveness of soils, especially parts that have direct contact with the ground, such as spikes, plates, scarifiers, supports, among many others called Ground Penetrating Tools (FPS). In the case of SPFs, in addition to rapid wear, there is a reduction in the productivity of the items, as they decrease in size and length, reducing their penetrating and pulling power. According to Barros & Melo (2006), they consider abrasive wear to occur between moving surfaces, under the action of a load, where the presence of hard parts in the bodies promotes physical interactions that deform and break the surface causing the removal of material from the surfaces of the soil penetration tools.

Another important factor, as reported by Komatsu in the magazine Manutenção & Tecnologia , issue 181, (2014) is that the main function of the wear materials used in buckets for excavators, backhoe loaders and wheel loaders is to protect the structural component, guaranteeing it a longer service life. These wear materials, also known as

Soil Penetration Factor (SPF), improve the performance of the bucket on the various work fronts, guaranteeing easier material entry and, consequently, increasing the productivity of the equipment.

The FPS caster becomes more agile, allowing better utilisation of the steel used in the tips and guaranteeing a higher level of sharpness until the end of its useful life. In other words, it is possible to ensure greater productivity for the equipment, combined with a lower cost per tonne produced.

It's also worth explaining that rotation is the action of changing the position of the tips in the same bucket in order to even out the consumption of wear material. After all, excavators, backhoes and wheel loaders tend to attack the material from the side, wearing down the tips more than the middle of the buckets. Turning, on the other hand, is the movement of the SPF by 180 degrees at the ends, to optimise the use of each SPF.

Both swivelling and castor can be applied to excavator tips. In the case of loaders, due to the profile of the tips and the type of operation of the equipment, swivelling is more recommended.

Obviously the operation of the equipment has a significant influence on the useful life of the wear materials. In this respect, it is worth emphasising that the constitution of the FPS - in terms of the rigidity, composition and heat treatment of the steel alloy used in its manufacture - also makes all the difference to durability. The recommendation, therefore, is to look for suitable materials for each operation, assessing the hardness, abrasiveness and other characteristics of the material to be handled.

For excavators, for example, there are specific teeth for general application, for extra-abrasive material and for highly abrasive rock. In these cases, there are pointed materials, double-ended drills and "shovel" drills, which are more suitable for levelling, clearing and grading land.

The tines, as already mentioned, improve penetration into the soil and, to increase their productivity, there are technologies that are self-sharpening, i.e. they allow them to remain sharp as they wear down, keeping their performance at a high level.

CHAPTER 3

3.1 - PROJECT OBJECTIVES

3.1.1 - GENERAL OBJECTIVE

The general aim of this work is to analyse the types of large excavator tip profiles with and without hard material coating that exist on the mining market, and to develop a new tip profile that is more efficient, so that it can better meet operational needs in addition to cost-effectiveness.

3.1.2 - SPECIFIC OBJECTIVES

a) Defining the best tip profile for the excavators used in mining operations to cater for the different types of soil excavated;

b) Increasing the useful life of excavator tips so that they can be replaced at the time scheduled for preventive maintenance;

c) Evaluate the cost/benefit of the changes so that the company makes a financial gain.

CHAPTER 4

4.1 - LITERATURE REVIEW

4.1.1 - MINING

It is a term that covers processes, activities and industries whose aim is to extract mineral substances from mineral deposits or masses. This can include the exploitation of oil and natural gas and even water. As an industrial activity, mining is indispensable for maintaining the standard of living and advancement of the modern societies in which we live (Wikipedia, 2014).

According to the international classification adopted by the United Nations Organisation (UNO), mining is defined as the extraction, preparation and processing of minerals in their natural state: solid, such as coal and others; liquid, such as crude oil; and gaseous, such as natural gas. In this broader sense, it includes the exploitation of underground and surface mines, quarries and wells, including all complementary activities to prepare and process minerals in general, in order to make them marketable, without irreversibly altering their primary condition.

4.1.2 - MINING PROCESS

Mining and processing operations involve a wide variety of stages, each with its own attributes and requirements to increase efficiency. The conditions for improving one of these stages can be counter-productive, leading to a drop in the performance of another stage. In other words, from the extraction of the ore to the final processing, the process has several separate areas that are dependent on each other.

Mining operations in an open-cast mine basically comprise four unit operations (mine process): drilling, blasting, loading and transport. Then there are the ore beneficiation processes (plant process): primary crushing and for beneficiation (plant process) we have primary crushing, secondary crushing, screening, long-distance conveyors, jigging, flotation, among other processes for the final products, such as *Pellet Feed, Sinter Feed, Coarse Granules, Fine Granules and Pellets.* The large number of stages in mining and processing operations, as well as the complexity and interrelationships between them, make trial and error processes difficult and expensive.

The object of study in this research is within this process, in the loading part, which is done by excavators and/or wheel loaders.

4.1.3 - OPEN-CAST MINING PROCESS

According to Wikipedia, 2014, Open pit mining refers to the method of extracting rocks or minerals from the earth by removing them from an open pit or borrow excavation. The term is used to differentiate this form of mining from extractive methods that require tunnelling into the earth - underground mining. Open-pit mining is used when deposits of commercially useful minerals or rocks are found close to the surface; that is, where the thickness of the ground cover (located above the material of interest, and which has to be removed to reach it) is relatively small or the material of interest is structurally unsuitable for tunnelling (as is the case with sands, volcanic ash and gravels). Where minerals occur far below the surface, and the thickness of the cover land is great or the mineral occurs in veins in the rock - the material of interest is extracted using underground mining methods. Open-cast mines are typically expanded until the mineral resource (or the plot of land owned by the mining company) is exhausted.

In open-cast mines, mining planning is fundamental for predicting the final pit, the mining sequence, the location of the fines containment basins, the controlled disposal of waste rock, the development of new products and permanent access projects (Queiroz Filho et al., 1997).

According to Casquet et al. (1996) and Queiroz Filho et al. (1997), apud Pinheiro (2000), the basic objectives of planning and carrying out iron ore mining are:

a) To guarantee the continuity of ore extraction operations in a mine so that the mine will always supply ore until the mine in question is finally exhausted;
b) Ensure the transport of waste rock and ore within a mine over long, medium and short-term periods;
c) To develop the mine safely in all respects for mine and plant equipment and for people;
d) Minimising environmental impacts, mainly by eliminating oil leaks from machinery and extracting only in areas cleared for the mining process;
e) Obtaining the appropriate blending (mixing of materials) of the ores to meet the quality requested by the clients;
f) Meeting production, quality and maintenance targets;
g) Achieve higher product quality;
h) Maximising the useful life of the deposit, ensuring greater equality of products and seeking sustainability;
i) Subsidising the economic exploitation of ore, planning from the removal of the ore from the mine to its arrival at any customer anywhere on the planet;
j) Achieve minimum cost conditions.

Figure 4.1 shows some open-cast iron ore mines.

Figure 4.1 - Examples of open-cast iron ore mines.

Source: *http://www.ebah.com.br/content/ABAAABBn4AL/lavra-mina-ceu-aberto-subterrânea*

According to Guerra (1988), methods for evaluating reserves can be classified into three main groups:

a) Conventional methods: these allow mineral reserves to be calculated using weighted average factors (content, thickness, inverse power of distance method and volume), which are applied to areas or volumes of influence;
b) Statistical methods: takes into account the spatial aspect of the area or volume of influence of a sample. Samples must be chosen randomly within the deposit. These are purely probabilistic methods that consider the geological process to be totally random;
c) Geostatistical methods: arose to take into account the spatial correlations between samples, as well as the randomness represented by unforeseen variations from one point to another in the deposit.

Guerra's (1988) proposal presents statistical methods for evaluating reserves, based on random sampling carried out on the mineral deposit.

Quevedo (2009) also considers the three basic methods for evaluating mineral reserves. Of these three, he reports that geostatistical methods are the most widely used for evaluating mineral deposits, as they take into account structural aspects and consider the randomness characteristic of mineralised formations.

Once the mineral deposit has been assessed, it is divided into geological blocks and the geological database of the deposit is assembled. After this stage, the mining project can be drawn up.

When drawing up the mining project, a study is carried out to size the equipment and installations that will operate in the mine, based on the production determined.

These fleets of equipment can be divided into five main classes, according to Quevedo (2009):

a) Drilling Equipment: includes the drilling rigs responsible for drilling to place blasting materials and shaking the ground;
b) Blasting equipment: this is represented by the tractors that are responsible for opening up accesses (roads) in the mine, as well as clearing tracks and stockpile areas;
c) Loading equipment: we have loaders and excavators, which can be hydraulic or electric. This equipment is responsible for loading material (ore and tailings) onto lorries for transport, with the ore going to ore treatment plants and the tailings to tailings piles;
d) Transport equipment: we have road or off-road trucks, which are responsible for transporting materials within the mine to the ore treatment facilities or waste rock pile.
e) Support Equipment: responsible for the tracks and inputs. This includes graders, water trucks, small loaders/excavators, supply trains and all the equipment that supports mining activities;

According to Pinto, T. P. (1999), when choosing the type and size of equipment, different factors must be taken into account, such as the scale of production, the financial capacity of the mining group and the characteristics of the mine, testing the various alternatives available.

There are mines, for example, that use a mix of loaders and lorries, with belts that transport the ore from the mining front to the crusher. Other mines use semi-mobile crushers, i.e. the position of the crusher is changed from time to time within the mine, so that the crusher is as close as possible to the mining fronts, reducing the transport distance and the material is already transported after a crushing,

The final sizing of the equipment only takes place once the mine's mining plan has been drawn up, as this gives an idea of the mine's geometry, as well as the accesses to each extraction point.

The geometry of the mine, as well as the material to be removed, is determined through a mining plan, which can be short-term (up to one year of planning), medium-term (one to five years of planning) or long-term (more than five years of planning). A mining plan can be considered to be a roadmap for the mine's operations within a certain timeframe that fulfils, on average, the requirements dictated by the treatment plant.

According to the Mining Regulatory Standard (NRM) (2002), drawn up by the National Department of Mineral Production (DNPM), the geometry of the pit, stockpiles and other structures must be updated every six months or more frequently, at the discretion of the DNPM, in accordance with the rate of progress laid down in the Mining Plan, which must be kept at the mine, as well as the relevant topographical documentation, for examination by the inspectorate.

The mine's geological control plans should be updated every six months, frequently reviewing all aspects related to the stability of the structures.

The project must have a collection of plans that, as a whole, cover the following items, as appropriate, according to Norma Reguladora de Mineração (NRM), (2002):

a) The limits of concessions;
b) Pit perimeters and disposal systems;
c) Safety lane boundaries;
d) Side angles of safety lanes;
e) Limits of the mining area;
f) Data relating to the thickness of the ore or layers mined;
g) The geological contacts of the different cuts in the cover and ore;
h) Elevations at significant points such as the upper and lower limits of the cuts in the cover and ore, at distances of less than 200.00 m (two hundred metres);
i) Revegetated areas;
j) Intercepted faults and dykes;
k) Delimitation of areas of risk and influence of mining.

A mining plan can be considered to be a roadmap of the mine's operations within a given timeframe that meets, on average, the requirements dictated by the treatment plant.

4.1.4 - LOADING EQUIPMENT

In a mining operation there are various types of products being excavated, as can be seen in Figures 4.2, 4.3, 4.4 and 4.5. These products have to be extracted from one mining front, but then take different paths afterwards. There is ore, waste rock, blocks and so on, each with its own chemical composition and especially mineral characteristics such as hardness, resistance to impact, friction and so on. When excavating, the excavator has to work with this variety of materials and characteristics.

Figura 4.2 - Mining front with ore and tailings.

Source: Mineração Usiminas (Mina Oeste - Itatiaiuçu - 2013)

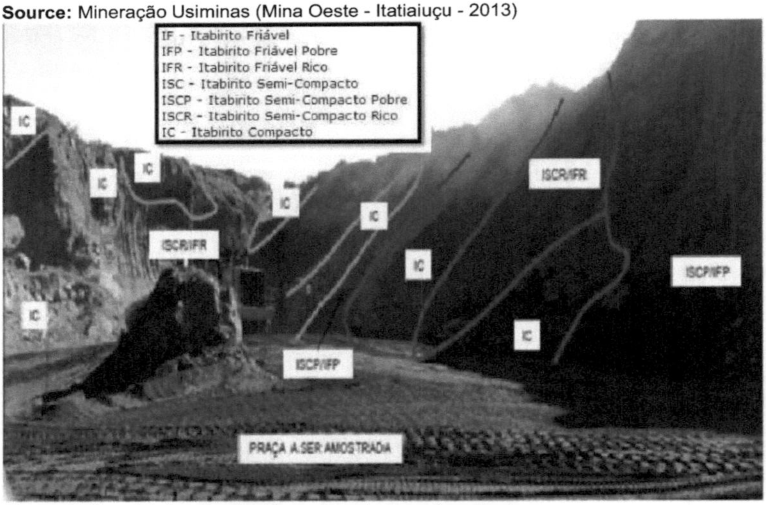

Figura 4.3 - Mining front with various types of products.

Source: Mineração Usiminas (Mina Oeste - Itatiaiuçu - 2013)

Figure 4.4 - Mining front with various rocks and compact materials.
Source: Mineração Usiminas (Mina Oeste - Itatiaiuçu - 2013)

Figure 4.5 - Mining front.
Source: Mineração Usiminas (Mina Oeste - Itatiaiuçu - 2013)

According to Quevedo (2009), loading and transport operations consist of transporting the material extracted from the quarry to different unloading points.

The definition of soil penetration tools (FPS) is directly linked to the use to which the equipment will be put, such as moving soil or rock, and to its operational characteristics, such as torque and power. When used improperly, accessories such as tips, edges, corners, segments, monobloc teeth, adapters, supports and nails can wear out, significantly compromising the equipment's performance and productivity (Revista de Manutenção & Tecnologia, Issue 147, 2011).

In open-cast mines, activities begin with the preparation of the area to be mined so that it can be drilled and blasted. The trucks are then directed to a specific mining front, and the loading equipment (wheel loaders or excavators) that is allocated to the fronts removes the material and loads it onto the trucks. The loaded lorries transport the material to a specific unloading point (crushers, waste pile or lung pile) and then return to an available mining front, where they repeat the same operations.

Loading and transport operations are carried out continuously. The lorries carry out loading and tipping cycles repeatedly along the possible routes available; when they leave a loading point for a tipping point, or vice versa, they do so directly without intermediate stops.

Figure 4.6 shows a schematic diagram of how lorries are handled in loading and transport operations. Loading operations on road or off-road lorries can be carried out with excavators or loaders. Figure 4.7 shows a large excavator loading an off-road lorry.

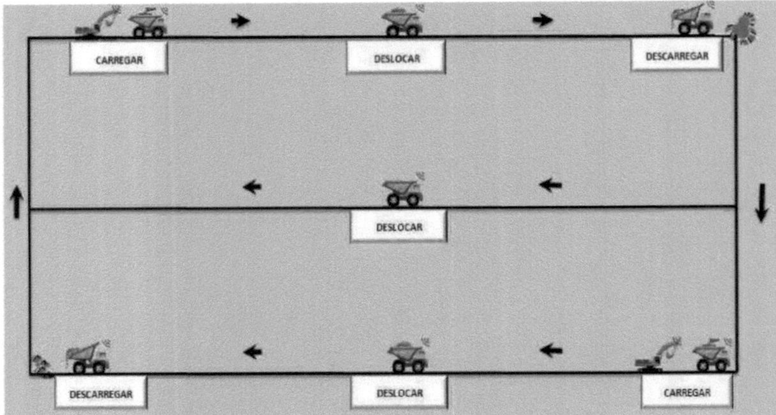

Figure 4.6 - Loading and transport process in an open-cast mine.
Source: Author

Figura 4.7 - Loading point with an excavator loading an off-road lorry.

Source: http://pt.dreamstime.com/imagem-de-stock-máquina-escavadora-que-enche-o- truck-image31565251

4.2 - SOIL PENETRATION TOOL - FPS

Ground penetrating tools are known by many names, some of which are: cutting edges, teeth, guards, wear iron, spikes or ground penetrating tools (FPS).

Whatever you call this tool, it is always a sacrificial component. Its purpose is to protect high-cost components such as the base edge of buckets, the structural plate ("mother" plate) and all the equipment structures that come into direct contact with the extraction of ore, waste rock, rocks and other materials on the benches and floors.

SPFs contribute a large part of a machine's maintenance costs. Although it is essentially a wear part, it is necessary to check which SPF is most suitable for each type of work in order to obtain the maximum useful life under wear conditions and reduce machine downtime for replacement.

In the quest for better equipment performance and in the face of requirements such as efficiency during penetration into the pile of material, abrasion resistance and impact resistance, users need to evaluate each case in order to make the best choice of these components. There are a number of companies involved in the manufacture of FPS and wear parts that point to a trend towards the use of increasingly thin tools.

To ensure the desired efficiency, specialists are using sharper points that provide rapid penetration in various types of soil, especially in soils such as gravel and low-abrasive ores. Tools with this profile break down the material more easily, providing faster digging cycles. The only caveat that experts report is that thinner tips are less durable.

To avoid breaking the FPS and constantly stopping to change it, some users prefer to use reinforced tools, i.e. tools with thicker tips.

Today, thicker tips are recommended for extremely heavy-duty work such as loading rock, limestone, granite and iron ore, i.e. in situations where the material's ability to break down is not an obstacle to productivity.

The FPS are mounted on buckets and other components using latches, pins or are welded directly to the "mother plate". Tips and shanks are mounted. The rest are welded directly to the components.

The most suitable materials for applications in wear, penetration and material handling parts are steels with alloys based on carbon, silicon, nickel, manganese, molybdenum and chrome, which can vary the mechanical characteristics of the assembly, as they are more resistant to abrasion and impact conditions.

Figures 4.8 and 4.9 show the ground-penetrating tools used in buckets, as well as the wear materials that also have direct contact with the ground, preventing wear of the structural plate in buckets that have contact with more abrasive wear materials. Item 2 in Figure 4.8 is the subject of this study.

Table 4.1 details each of the wear components installed in a bucket on the structural side, with their functions. This bucket is very similar to the bucket used in the study.

Figura 4.8 - Front detailing of the bucket with SPF and wear materials.

Source: http://www.excavatorbucket.en.alibaba.eom/product/1763067166-221940381 / Kobelco SK350 Underground Mining Bucket for Excavator.html

Figura 4.9 - Bucket side detailing with SPF and wear materials.

Source: http://www.excavatorbucket.en.alibaba.eom/product/1763067166-221940381 / Kobelco SK350 Underground Mining Bucket for Excavator.html

Table 4.1 - Ground-penetrating tools and wear materials.

Number	Component	Function
1	Cinnamon	Side protection - fixed with pins and/or by welding.
2	Tips	First part of the bucket that touches the ground - fixed with pins or locks.
3	Internal coating	Internal side protection - fixed by welding on the sides and riveted in the centres.
4	Between teeth	Protection between the teeth - fixed by welding.
5	Internal bulge lining	Internal bulge protection - fixed by welding.
6	Corner	Edge protection - fixed by the welding process.
7	External lining	Rear bucket lining - fixed by welding process
8	Bulge side panelling	Lining the sides of the bucket - fixed by welding.

9	Handle	Bucket fixing bracket - fixed using the welding process.
10	Wear pads	Reduce material contact with the side of the bucket - fixed by welding.

Source: Author

4.3 - HYDRAULIC EXCAVATORS IN MINING

Also called wheel loaders, excavators or mechanical shovels, excavators are the generic designation given to various types of machines for digging, turning or removing earth or any material that needs to be removed, such as: earth, concrete blocks, rocks, tailings ponds, iron ore ponds and other metals found in nature.

Excavators can carry out various operations, depending on the type of boom.

a) front shovel boom with inverted bucket;

b) *front-shovel* boom with *shovel bucket-,* c) boom with *drag* bucket or *drag-hne* d) boom with jaw bucket or *clam-shell-*, e) boom with multi-joint bucket or *orange peel-,* f) *backhoe, back-shovel, retro-shovel* or *hoe* boom.

The first two buckets are part of the study. The others have no or little use of ground-penetrating tools.

The front bucket excavator is also known as a backhoe loader with an inverted bucket, as it has the bucket facing downwards and digs by moving from top to bottom, as shown in Figure 4.10. This implement is most efficient when digging at a lower level than the base support. The operation of the backhoe loader is similar to that of the shovel, but differs in terms of unloading the bucket. Loading is done through the bucket mouth and unloading is also done through the bucket mouth.

The front bucket excavator with shovel is a self-propelled machine fitted with an articulated boom (also called a tower in some publications), with an equally articulated arm and a mobile bottom bucket at the end. The bucket faces upwards and digs by moving from bottom to top, as shown in Figure 4.11.

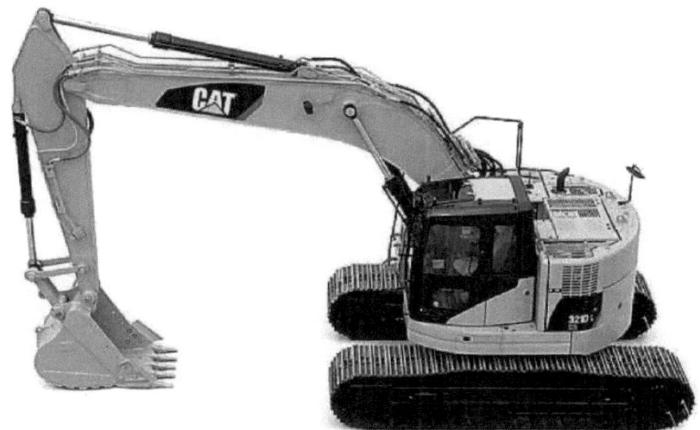

Figura 4.10 - Excavator with inverted bucket.
Source: Caterpilar Model 321D - 2014

Figura 4.11 - Excavator with shovel bucket.

Source: Caterpilar Model 6090 - 2014

Showel bucket excavators are used to excavate high cuts, especially rocky materials, with slopes located above the level at which the machine is working (Ricardo and Catalani, 2007; Abram and Rocha, 2000; Nunnally, 2011; Peurifoy et al., 2011). The rotation of the boom allows the bucket to be moved in the horizontal plane to an unloading position which is carried out by opening the bottom of the bucket. The shovel is the ideal piece of equipment to be used in "heavy services" due to the high digging force obtained at the cutting edge of the bucket and the safety it offers.

The *drag-line* bucket scaffold, on the other hand, *is* made up of a metal lattice at the end of which there is a pulley through which the bucket lifting cable passes, driven by the capstan. The boom is supported by the cable, varying its angle between 25° and 40° by means of a hinge, and its radius can be increased by adding an intermediate section. Excavation takes place by dragging the bucket via the drag line, which is driven by the capstan. *Drag-line* excavators work in excavations at levels below those at which they are located (Ricardo and Catalani, 2007; Abram and Rocha, 2000; Nunnally, 2011; Peurifoy et al., 2011). The advantages of the *drag-line* are:

a) Longest reach of all earthmoving equipment;
b) The possibility of loading transport units outside the excavation area, preventing them from having to manoeuvre over mud.

The *clam-shell* boom consists of two moving parts, controlled by cables that can open or close like jaws and have a cutting surface or teeth. Excavation takes place by the bucket falling and then the jaws closing, so that the excavation progresses vertically.

The excavator with a backhoe bucket has the bucket facing downwards, towards the cab of the machine, and for this reason it works by excavating at levels below the one at which the machine is located. According to Ricardo and Catalani (2007), this type of excavator is used:

a) when digging trenches of great depth and small width, without shoring;
b) on high cuts;
c) as a substitute for *drag-line* excavators when opening canals, removing unsuitable soils, etc. In Brazil, this type of equipment is simply called a hydraulic excavator. There is also the so-called backhoe *loader*, known for its versatility, which consists of a combination of three types of equipment - tractor, loader and excavator with backhoe boom (Peurifoy et aL, 2011).

Hydraulic excavators are generally divided into three sizes: mini/small, medium and large. Table 4.2 shows the capacity of each excavator according to its size.

Table 4.2 - Hydraulic excavator capacities.

Size	Operating weight (tonne)	Power (kw)
Mini / Small	7 a 20	59 a 93
Medium	21 a 37	103 a 200
Big	Greater than 37	Largest 200

Source: Caterpilar (2014)

The largest excavator in the world was built by the German company Krupp, and is 96 metres high and weighs 13,500 tonnes. Figure 4.12 shows a model of a large excavator, highlighted in white (light) and a small excavator (mini) highlighted in orange (dark) which is inside the bucket of the large excavator.

Figure 4.12 - Large Showel excavator and mini inverted excavator.

Source: http://www.asmaquinaspesadas.com/2012/06/fotos-e-videos-da-maior-escavadeira ▪html

Experts in the sector warn that the type of tool suitable for each operation must be considered according to the service it will perform, especially in relation to the geometry of the tips and even the blades (cutting edges) of the buckets (Revista de Manutenção & Tecnologia, Issue 147, 2011).

The correct specification of ground penetrating tools (FPS) also results in lower fuel consumption, less strain on the equipment and helps to reduce maintenance and operating costs (Revista de Manutenção & Tecnologia, Issue 147, 2011).

In general, the PSFs of crawler tractors, motor graders and backhoe loaders have similar designs , with little variation. The greatest variations in terms of PSF tip geometry are found in loaders and excavators and are highlighted within this study.

Figure 4.13 shows a view of a common hydraulic excavator on the market made up of 29 components. Table 4.3 details each of the components with their respective functions.

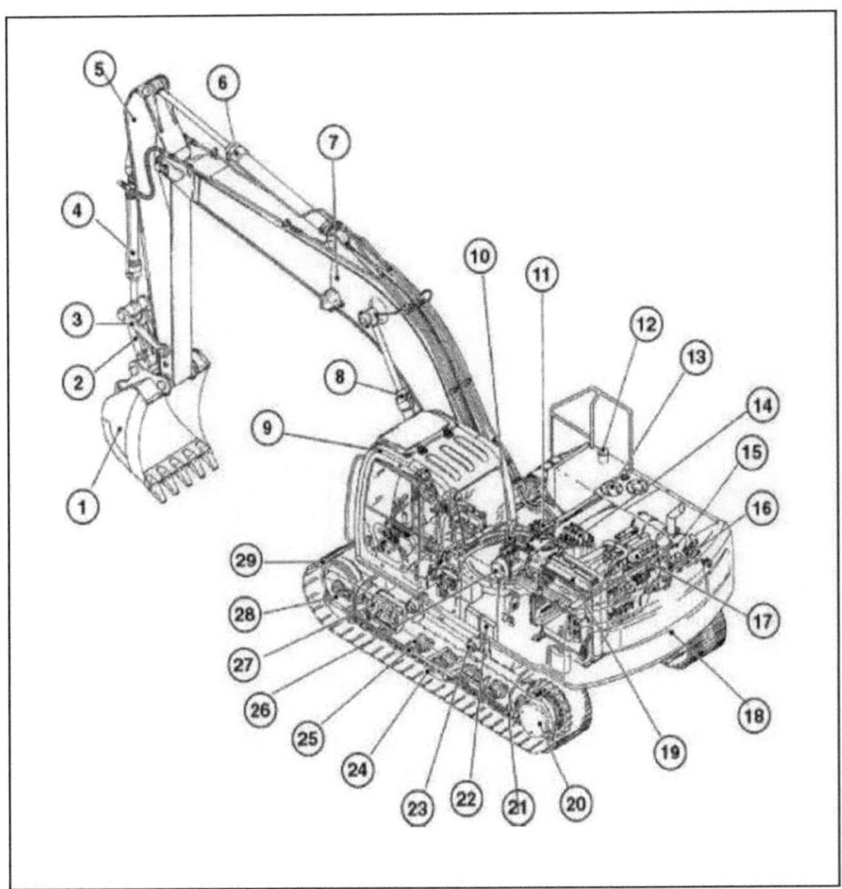

Figura 4.13 - Components of a hydraulic excavator.

Source: http://c3manuais.blogspot.com.br/2012/10/curso-escavadeira-hidraulica-case-cx.html

Table 4.3 - Breakdown of the components of a hydraulic excavator.

Number	Component	Function
1	Bucket	Responsible for removing material from the ground. Contains the tips.
2	Articulation of the bucket	Responsible for closing and opening the bucket.
3	Intermediate articulation	Responsible for closing and opening the bucket.
4	Bucket cylinder	Responsible for closing and opening the bucket.
5	Penetrator arm	Responsible for closing and opening the bucket.
6	Arm cylinder	Responsible for the complete movement of the arm.
7	Launch	Responsible for the complete movement of the arm.

8	Boom cylinder	Responsible for the complete movement of the arm.
9	Cabin	Place where the machine operator moves the equipment.
10	Speed reducer	Responsible for turning the equipment 360° in both directions of rotation.
11	Turning motor	Responsible for turning the equipment 360° in both directions of rotation.
12	Reservoir Fuel	The equipment's fuel tank.
13	Hydraulic oil reservoir	Hydraulic oil tank for the entire hydraulic system.
14	Control valve	Hydraulic valve that directs the flow of oil to all the equipment's movements.
15	Engine silencer	Responsible for reducing noise and pollution from the gases expelled by the diesel engine.
16	Hydraulic pump	Responsible for pumping hydraulic oil, generating calibrated pressure and flow for the hydraulic system.
17	Diesel engine	Diesel fuelled internal combustion engine.
18	Counterweight	Responsible for levelling the machine when it is in loading operation.
19	Engine radiator	Component responsible for controlling the temperature of the diesel engine via a cooling water system.
20	Travelling motor	Responsible for moving the machine in both directions (forwards and backwards). It has a toothed crown mounted on the slewing motor.
21	Treadmill	It works like a bicycle chain and is mounted on the slewing ring on one side and supports that rest on the floor and move the machine.
22	Electrical system	Made up of batteries and electrical components.
23	Top roller	Helps to turn the treadmill.
24	Treadmill guide	Responsible for centralising the conveyor belt.
25	Bottom roller	Helps to turn the treadmill.
26	Lubrication pump	Responsible for sending grease to the main mobile joints.
27	Tensioning spring	Responsible for tensioning the belt.
28	Guide wheel	Helps to turn the treadmill.
29	Treadmill	Contact with the ground moves the equipment in both directions.

Source: Prepared by the author

The movements of the hydraulic excavators are controlled by a system of positive oil pumps that provide the movements of travelling on tracks, rotating 360° to both sides (right and left) as well as driving the hydraulic "arm" that does all the loading of material onto the transport trucks.
The machine's movements copy those of the human arm, where the boom and sf/ck are respectively the human "arm" and "forearm", while the bucket is like the "hand" carrying the material.

A hydraulic excavator is made up of three pieces of equipment: a tractor, a loader and an excavator. Each of these pieces of equipment is suitable for a particular type of work, and generally all three pieces of equipment are used on a construction site.

The tractor is designed to move over all types of rough terrain, it is the central structure of the excavator and most machines are made up of metal tracks.

The loader, which is the front part of the excavator, is mainly used to load earth, dirt and other materials in large quantities. It is also used to flatten or push them like a plough.

Figures 4.14 and 4.15 show, in schematic form, the possible movements of inverted bucket excavators and shovel bucket excavators respectively. It can be seen that the bucket can pick up above and below the equipment itself. The equipment can rotate 360° to either side, which makes it possible to remove material within a maximum radius of the excavator. Another feature of the equipment is that it moves forwards and backwards at the same speed and force.

Finally, the main tool is the excavator itself, which is used to compact material, lift heavy objects and excavate. It digs all kinds of holes, but its main function is to dig trenches.

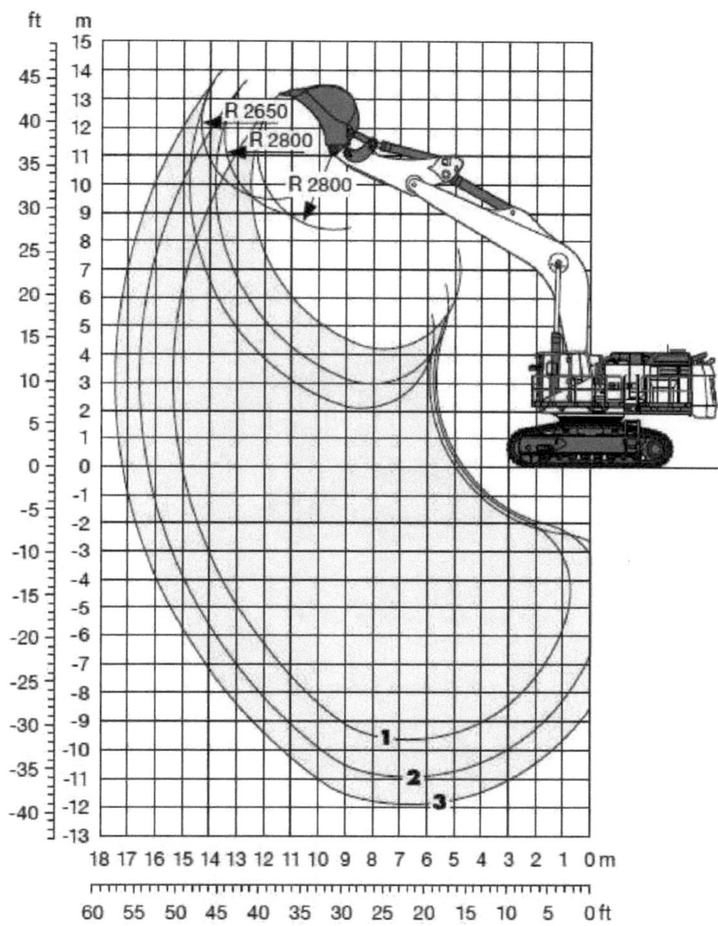

Figura 4.14 - Possible movements of an inverted bucket hydraulic excavator.

Source: Liebherr Model R9100 (2014)

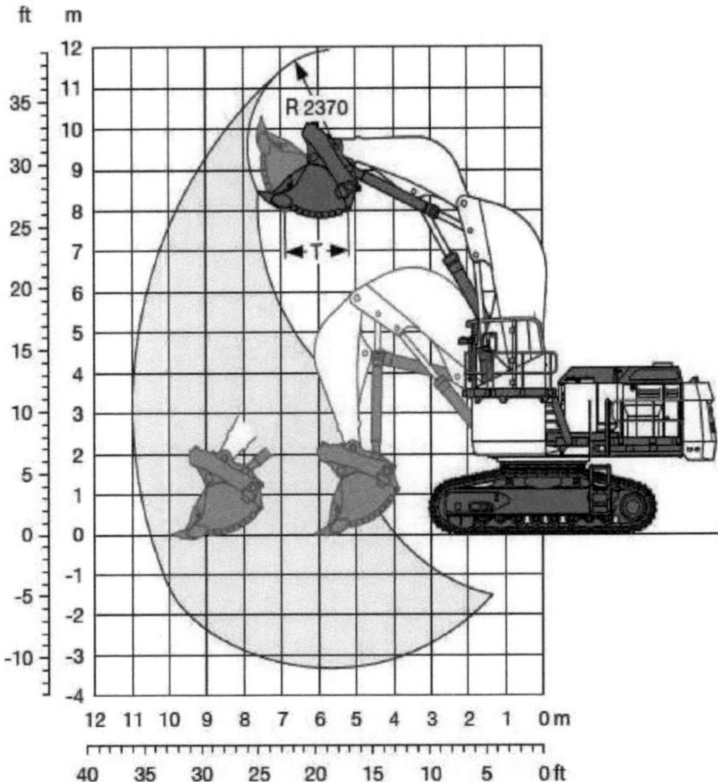

Figura 4.15 - Possible movements of a shovel-type hydraulic excavator.

Source: Liebherr Model R9100 (2014)

Figure 4.16 shows a model of an excavator that is used in mining and is the machine that was defined for this study. This excavator weighs 110 tonnes and is equipped with a 565 kW (757 hp) engine. Its bucket can carry up to 15 tonnes in a single "scoop" or "dig".

Figura 4.16 - Model of the hydraulic excavator used to carry out this work.

Source: Liebherr R9100 (2014)

4.3.1 - EXCAVATOR BUCKET DETAILS

According to the technical specifications of machine suppliers such as *Caterpilar, Liebherr, New Holland and* others, the bucket is the general-purpose component used in hydraulic and cable excavators of any size and capacity, and is a fundamental accessory for carrying out excavation work in particular.

Buckets are usually the part of heavy equipment that has direct contact with earth, sand or rock, whether they are loaders, backhoes, hydraulic excavators or dump trucks. In heavy machinery, buckets are used for excavating and transporting the excavated material or loading it.

In machines, buckets can be common with blades, teeth or nails. They can also be crusher buckets, for crushing concrete, rock or rubble, or sifter buckets, for gradually separating materials, among many other types.

The volume of excavator buckets depends on the buyer's needs. The characteristics of the buckets will depend on their size, the type of work to be carried out, the lorries that will be loaded, as well as other technical details that must be informed to the manufacturer when purchasing the bucket.

The volume and density of the material to be transported determines the size of the bucket. But the main characteristic that will determine what the bucket is like and whether it will need reinforcements or wear materials is the soil on which the excavator will operate.

An excavator that works with "soft", loose soil has less effort than one that works excavating iron ore benches (high abrasiveness) or ripping up and moving rocks (multiple impacts on the tips and sides). This also has a direct impact on their wear materials and ground-penetrating tools.

For example, the excavator bucket for rock requires certain characteristics that provide greater resistance to wear from abrasion or impact, which is why they are made from high-strength structural steel.

Figure 4.17 shows in detail the parts that make up an excavator bucket. This bucket is made up of structural plates, also known as the "mother plate", wear plates on all its surfaces that have contact with the ground to prevent wear on the structural plates and, above all, the ground penetrating tools (FPS), which is the tip. The structural plates should have as little contact with the ground as possible to avoid wear that would lead to the manufacture of a new bucket.

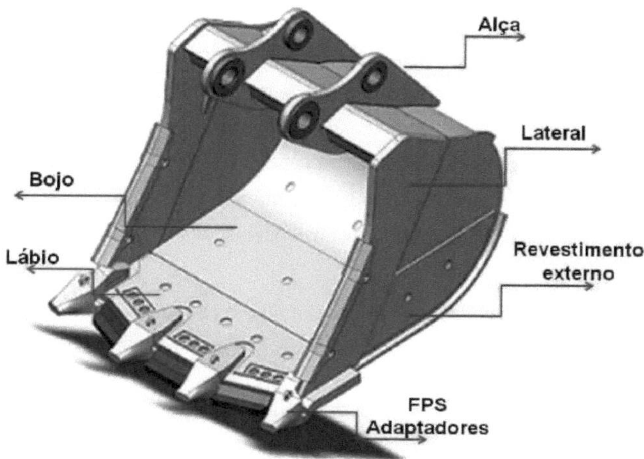

Figura 4.17 - Detailing the bucket components of a hydraulic excavator.

Source: Author

According to the technical specifications of the bucket supplier in the study, the materials used in the bucket assemblies are described in Table 4.4.

Table 4.4 - Breakdown of bucket components.

Component	Function	Material
Handle	Bucket frame plate that connects to the excavator.	USI-SAR-80
Side	Reinforced plates on the sides for contact with the ground.	USI-SAR-60
Coating External	Bucket lining on the sides and behind the bucket in areas that have direct contact with the ground.	USI-AR-450
Lip or Edge	The front of the bucket, after the tips, is the second area of contact with the ground and wears out due to abrasion.	USI-AR-450
Adapter	Ground-penetrating tool that is mounted to the wear tips that are the bucket's first contact with the ground.	USI-AR-450
Bulge	The internal part of the skip responsible for receiving the entire volume of material. It can have wear material on the inside or outside.	USI-SAR-60

Source: Prepared by the author

According to the table of materials (Catalogue - Usiminas - Heavy Plate, 2014), the USI-SAR-60 and USI-SAR-80 structural steel grades involve ultra-high mechanical strength materials with guaranteed toughness at low temperatures and superior welding performance. They are produced by controlled rolling and accelerated cooling. Normalising or Quenching and Tempering. They are characterised by their low carbon equivalent, which gives this grade excellent weldability.

The USI-AR-450 material, on the other hand, is steel with added alloying elements, hardened, with high hardness as its main characteristic, and is intended for high mechanical wear services. This class includes materials with a Brinell hardness in the 360 to 550 HB range. These steels also have good weldability and, in special cases, on request, can be supplied with a *Charpy* impact guarantee at -20°C or lower. They are used in tractors, backhoes, off-road truck buckets, hoppers, chute linings, ore conveyors, blast furnace parts and industrial fans (Catalogue - Usiminas - Heavy Plate, 2014).

The edge or lip must have a minimum hardness of 345HB to 499HB, which remains practically constant from the surface to the core of the excavator bucket, i.e. there is no hardness degradation as wear occurs. The handles are also developed with greater mechanical strength, thus increasing the service life of the excavator bucket (Catalogue - Usiminas - Heavy Plate, 2014).

The material used to make the teeth, adapters and tips is a steel with variations of manganese (to increase the steel's mechanical strength), chromium (to increase abrasion resistance) and nickel molybdenum (to increase hardenability and toughness). The hardness of this material generally varies between 400 and 500 HB (Catalogue - Usiminas - Heavy Plate, 2014).

Other suppliers of buckets for backhoes, loaders and hydraulic excavators use materials with greater resistance to wear, such as HARDOX 450, which has a minimum hardness of 345HB to 499HB, which remains practically constant from surface to core, i.e. there is no degradation of hardness with wear. This type of bucket for backhoe loaders is used in quarries, mining companies, granite quarries and any other type of heavy duty work. For general uses such as earthworks and excavations, the construction of the bucket for backhoe loaders, wheel loaders and excavators is done without HARDOX plate reinforcement (Catalogue - Usiminas - Heavy Plate, 2014).

These buckets are used on various machine models such as: *Fiatallis, Case, New Holland, Michigan, Volvo, Caterpillar, Komatsu, Hyundai, John Deere, Massey.*

Figure 4.18 shows the bucket that was the subject of this study.

Figure 4.18 - Bucket on a large excavator.

Source: Liebherr R9100 Excavator Bucket (2014)

4.4 - TYPES OF WEAR

Wear is the progressive loss of substance from the surface of a body as a result of relative movement with the surface (Gahr, 1987). According to Gregolin (1990), wear has a direct impact on production costs due to the need to replace or recover worn parts, and also indirectly on production costs due to the need to oversize components and production limitations due to deteriorated equipment, as well as unforeseen interruptions in production lines.

According to Eyre (1978), wear can be defined as the degradation of the surface of a component or piece of equipment, involving the progressive removal of material as a result of tribological processes. He noted that abrasive wear is the most frequent form of occurrence in industrial segments, contributing around 50% of industrial problems involving wear.

Rigney (2009) defines wear as the displacement or removal of material resulting from tribological processes, while DIN 50 320 (1997) defines it as the progressive loss of substances from a solid body caused by mechanical action, i.e. by contact and relative movement of a solid body against another solid, liquid or gaseous body.

Due to the complexity of the factors involved in wear, classifications have been established to facilitate the study of the phenomenon and its prevention. According to DIN 50 320 (1997), wear comes in different types, the most common being: abrasive, erosive, friction, surface fatigue and tribochemical reaction wear.

In industrial environments, the wear found can be presented as (percentage of occurrence). According to Albertin (2003), the wear mechanism with the highest percentage of mass losses in practice is abrasive wear, which is around 50% of all other types of wear. This is followed by erosion with 15%, friction with 8% and corrosion with 5%.

It is also recognised that resistance to wear and tear is not intrinsic to a material, but rather to the characteristics of the systems or equipment to which the component is mechanically connected and its operating environment (Martins, 1995).

In the work carried out by Passos et al (2010), it was found that a *Komatsu* excavator, model PC4000, with a set of five tines, had to change 16 sets of new tines in just one month of work (480 hours worked), due to wear and tear, i.e. an average of 30 hours of work. These tines wore out due to the high abrasiveness of the soil, along with the machine's ability to penetrate it.

4.4.1 - FPS WEAR

According to an article published by Komatsu in issue 181 of Maintenance & Technology Magazine (2014), the main function of the wear materials used in excavator, backhoe loader and wheel loader buckets is to protect the structural component, guaranteeing it a longer service life. These components, also known as Ground Penetrating Tools (FPS), improve the bucket's performance on the various work fronts, ensuring easier material flow and consequently increasing the productivity of the equipment.

As the tillage tools move through the soil, various interactions occur due to friction between these two elements, causing the equipment to lose material and consequently wear out. This interaction causes changes in the geometric characteristics of the tools, which will result in greater tractive effort and difficulty in carrying out the work properly.

According to Machado et al (2009), although there are several studies on traction demand, little importance has been given to the wear of symmetrical tillage tools (tine tips), such as the work by

Mourad and Santos (2003), who built an equipment to assess tool wear depending on the type of soil in laboratory conditions, and Espírito Santo (2005), who assessed the wear of two tine tip materials of no-till seed drills in field conditions, due to direct contact with the soil.

Among the various forms of wear, the main one found in soil tillage is abrasive wear, which is the removal of material from the tool due to friction with soil particles. According to Mesquita and Barbosa (2005), the end of the tool's life is usually conditioned by wear, due to the high contact tensions associated with the relative sliding of the tool, and the process may contain high hardness particles in the sliding region, causing abrasive wear. According to Bhole and Yu (1992), abrasive wear best describes the removal of material from a solid surface by the action of the ground.

The frictional characteristics of soil on steel influence equipment performance in three ways: firstly, wear is a problem in both tillage and cultivation equipment due to the abrasive nature of many soils (Richardson, 1967); secondly, the force required to separate or move the soil not only depends on the physical properties of the soil, but also on soil/equipment friction (Hettiaratchi et al., 1966); finally, the magnitude of the friction of an internal nature of the soil establishes the degree of abrasiveness of the soil at the interface (Stafford & Tanner, 1977).

According to Baptista and Nascimento (2014), the wear of a tool working directly in the soil depends on several factors, including the soil's attributes and the tool's characteristics. According to Owsiak (1997), the wear of a soil cutting tool is at the mercy of soil conditions, operational factors and tool characteristics; he also points out that in research practice it is very difficult to cover all the influencing factors simultaneously and that, generally, only the effects of a few factors are investigated.

According to Fernandes et al. (2002), studying the wear mechanisms of the materials used in agricultural implements is fundamental for optimising the choice of materials and predicting the durability of equipment.

Improperly designed spikes lead to excessive replacements; this procedure becomes costly because, in addition to the number of tools used, it demands replacement time that could be used for effective work (Espírito Santo, 2005). It was found that the greater the tip wear, the greater the horizontal force required by this tool and, consequently, the more energy must be imposed on the equipment's traction and the tool's penetration into the soil.

The wear and tear of industrial, mining and agricultural components and equipment, as well as that of countless other industries, is a major factor in equipment depreciation and a source of expenditure on maintenance and replacement of mechanical components (Castro, 2010).

Wear directly increases the cost of worn parts and their replacement. It also often requires over-sizing components, limiting production due to worn-out equipment and drastically interrupting production. These factors have a significant influence on indirect losses in production output.

As wear is essentially a surface phenomenon, involving the undesirable mechanical removal of material from surfaces, the solutions found through weld overlays have proved highly valuable, both to prevent and to minimise or recover the different forms of metal wear. On the other hand, it makes production time much longer and can make coatings unfeasible and must be taken into account when calculating the cost benefit (Baptista and Nascimento, 2014).

In many situations, parts and components can be manufactured using conventional materials, within the normal design specifications. Subsequently, layers or weld seams can be applied to the surface, using suitable consumables to resist wear and tear.

Metallographic studies have shown that wear resistance is directly related to the microstructural characteristics of the material. Mourad and Santos, 2003 studied the wear resistance of heat-treated steels at various hardness levels and also of some pure metals. For pure metals, wear resistance increases linearly with hardness.

In ferrous materials, this relationship (wear resistance X hardness) is not simple. An increase in carbon content increases wear resistance. For steels with the same percentage of carbon, the wear resistance of an alloyed steel is higher than that of an unalloyed steel, but this increase is small when compared to one in which the percentage of carbon has increased (Baptista and Nascimento, 2014).

Wear resistance generally increases as the microstructure changes from ferrite to pearlite, from pearlite to bainite and finally from bainite to martensite, provided this is accompanied by an increase in hardness. However, for the same hardness value, the bainitic structure has greater wear resistance than the martensitic one (Baptista and Nascimento, 2014).

The microstructure has a greater influence on wear than the hardness of the matrix. The presence of retained austenite has been shown to improve the wear resistance of tempered martensite. Austenite provides better anchorage for the carbides, causing low carbide pullout from the austenitic matrix (Baptista and Nascimento, 2014).

Carbides seem to be particularly important in abrasion resistance, especially in materials such as chromium (Cr)-alloyed steels and white cast irons. Their influence is related to their hardness, size and distribution. Hard, finely dispersed carbides increase wear resistance, while coarse carbides decrease it.

In the specification and standardisation of wear-resistant alloys, the greatest difficulties lie in the lack of standardised tests to discriminate between levels of acceptance or rejection, depending on the particular applications.

These difficulties are mainly associated with the complex nature of the wear phenomenon. As well as involving surface deformation and cutting by abrasive particles, or friction between metallic surfaces, it often occurs due to several concomitant surface wear mechanisms, which can also be associated with other degradation phenomena such as impact, corrosion or fatigue (Dettogni, 2010).

Structures deformed by cold working do not increase resistance to wear, while increasing hardness by refining the grains would act favourably.

For ease of analysis and prevention, the predominant mechanism(s) of material removal are usually identified. To this end, the general types of wear can be categorised as (Baptista and Nascimento, 2014):

a) Abrasion wear - caused by abrasive (hard) particles under tension, travelling over the surface;
b) Erosion wear - due to the impact against the surface of solid particles or liquid droplets present in fluid currents;
c) Cavitation wear - associated with the formation and implosion of gaseous bubbles in fluid streams at the liquid-metal interface, due to sudden pressure variations along the way;

d) Adhesion or Friction Wear - resulting from metal-to-metal manufacturing, when rough surfaces slide against each other;
e) Corrosive Wear - which involves the occurrence of surface chemical reactions in the material, in addition to mechanical wear actions;
f) Impact wear - caused by shocks or loads applied vertically to the surface.

Figure 4.19 shows a bucket with ends already worn off during production. The wear is shown by the surface polishing on the tool set and on all the other wear plates welded to the bucket.

Figure 4.19 - Bucket of an excavator in operation.

Source: Vale (Mutuca Mine 2011)

Figures 4.20, 4.21 and 4.22 show the progressive wear of the tip assembly as the excavator is in operation. The tines operated for 160 hours in a friable soil (loose soil with low granulometry). The thin lines show the outline of a new tip in Figures 4.20 and 4.21. It can be seen that the tips are polished due to the friction of a fine material against them.

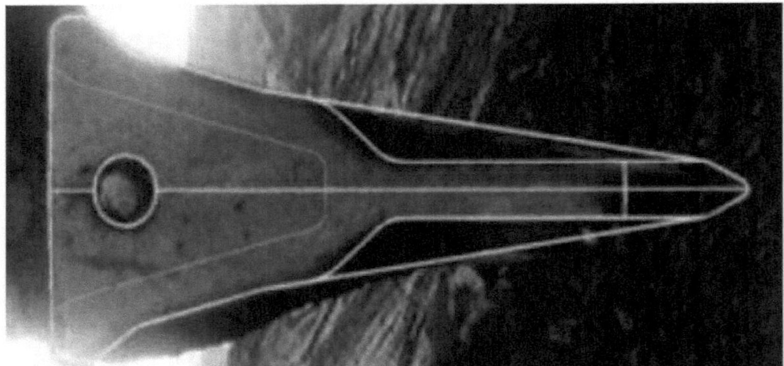

Figure 4.20 - Peak with 65 hours of operation.

Source: Vale (Mutuca Mine 2011)

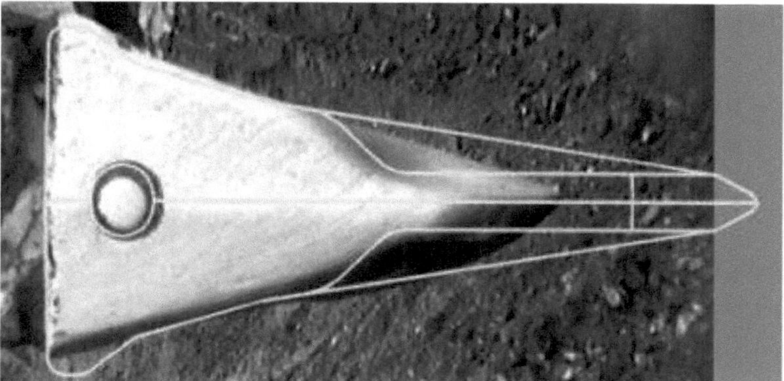

Figure 4.21 - Peak with 113 hours of operation.

Source: Vale (Mutuca Mine 2011)

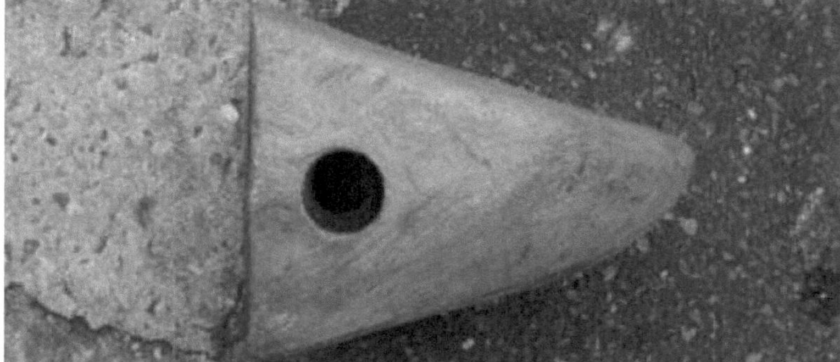

Figure 4.22 - Peak with 160 hours of operation.

Source: Vale (Mutuca Mine 2011)

Cladding welding is one of the largest fields of application for the prevention, minimisation and recovery of parts and components subject to abrasive wear, and forms of abrasive erosion, which are similar to abrasive wear (Baptista and Nascimento, 2014).

As abrasive wear is a complex phenomenon in which hard particles or asperities penetrate the surface of mechanical components, the vast majority of materials used in applications requiring high wear resistance are polyphase materials. These are usually made up of a hard phase with characteristics close to those of ceramic materials, surrounded by a ductile matrix (Baptista and Nascimento, 2014).

The study of these materials has shown that the size, distribution, hardness, ductility, toughness and volume fraction of the phases present are determining parameters in tribological development (Baptista and Nascimento, 2014).

4.5 - APPLICATIONS AND FORMS OF MINING EQUIPMENT FPS

Ground Penetrating Tools are applied to mining equipment. Figure 4.23 exemplifies this application in various operating situations and Figure 4.24 the main tools that have contact with the ground to be worked.

The shapes of excavator and loader tips are dimensioned by analysing three important points: tool wear material, penetration and impact.

The shape of the tip, as well as the reinforcement and the adoption of fastening systems that allow it to be replaced quickly, help to configure the most suitable tips for each application. In mining, excavator and loader tips have a short lifespan of around 80 to 300 hours, as the terrain is often abrasive, damaging them quickly.

Figure 4.23 - Mining equipment with its respective SPF.

Source: Vale (Various Mines 2011)

Figure 4.24 - Soil Penetration Tools.
Source: Vale (Equipment components 2011)

Figure 4.25 shows a guide in the form of bars to help choose the best tip depending on the type of soil to be worked. The longer the bar in the guide, the greater the characteristic. Tines are classified by three main characteristics, as described below:

a) Wear material - the longer the first (yellow) bar, the greater the wear resistance of the material to the ground;

b) Penetration - the greater the length of the second bar (grey), the greater the resistance of the material to penetration into the soil;

c) Impact - the longer the third (black) bar, the greater the material's resistance to impact with the ground.

Figura 4.25 - Tip selection guide for each type of soil.

Source: Sotreq (2014)

Figures 4.26 and 4.27 show the shapes of the tips related to the type of soil and the characteristics highlighted in the guide. They range from the simplest to the most detailed. As each of the characteristics increases, so does the cost of each tip.

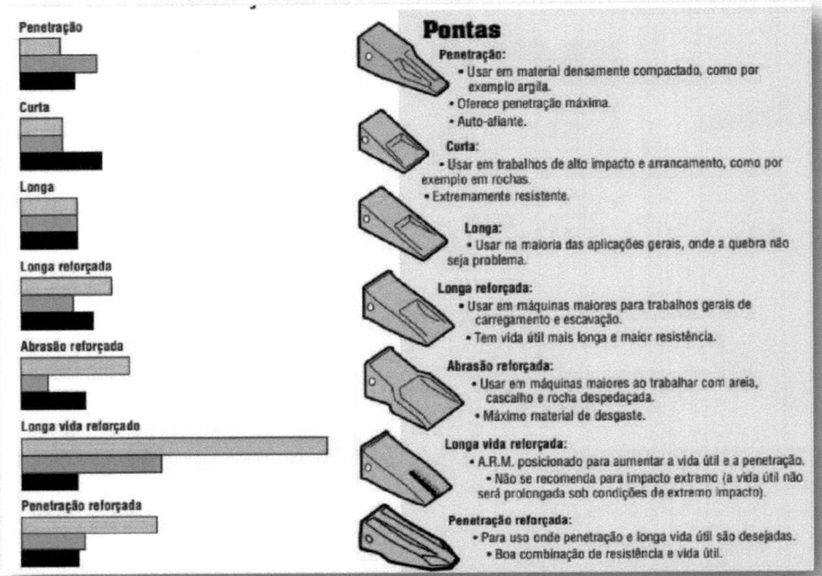

Figura 4.26 - Tips related to the type of soil.

Source: Sotreq (2014)

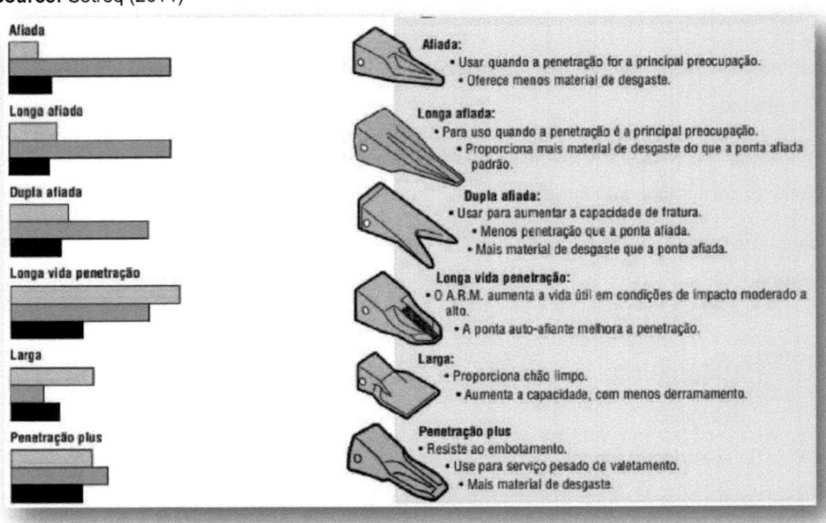

Figure 4.27 - Points related to soil type.
Source: Sotreq (2014)

4.6 - MATERIALS USED IN SOIL PENETRATION TOOLS

In order to meet specific applications such as highly abrasive environments, certain quantities of alloying elements different from those normally used in ordinary steels are added, forming alloy steels.

These quantities of alloying elements are defined with the aim of promoting changes in the physical and mechanical properties of the steel that allow the material to fulfil specific functions for each application. Each product has physical, chemical and mechanical characteristics.

As specified in the 2014 Purchasing Guide - Brazilian Steel and Non-Ferrous Metals, page 121, the main alloying elements added with their respective influence on the physical and mechanical properties of steel are:

a) Carbon - C - Melting point 3,650°C. It is the main alloying element in steel. By definition, "Steel is an iron-carbon alloy containing up to 2.11 per cent by weight of carbon." In steel, carbon is mixed with iron or in the form of harder carbides with iron, forming cementite or FesC, or with other elements such as chromium, molybdenum or vanadium. Therefore, it can be said that the main property conferred on steel by carbon is hardness. It also increases tensile strength and hardenability, but decreases toughness and weldability.

b) Chromium - Cr - Melting point 1,860°C. Element that favours the formation of carbides in steel. It therefore increases the hardness and tensile strength of the steel. It also increases hardenability and resistance to corrosion, including atmospheric corrosion, but slightly reduces toughness and greatly reduces weldability. On average, the tensile strength limit increases by 8 to 10kg/mm^2 with the addition of 1% chromium, but impact resistance decreases. It is the main alloying element in ordinary stainless steel when used at levels above 11%.

c) Manganese - Mn - Melting point 1,244°C. Increases hardenability, weldability and tensile strength. A stabilising element for austenite, it also alters the steel's transformation temperature, making it possible to obtain grain refinement and improved toughness during hot forming. Under specific processing conditions, it helps to generate a banded structure. It combines with sulphur to form manganese sulphide (MnS) which elongates during plastic forming and, if it occurs in large quantities, generates brittleness in the final steel. In the presence of carbon, higher levels of manganese greatly increase abrasion resistance, which is the basis of Hadfield steels.

d) Molybdenum - Mo - Melting point 2,620°C. Increases hot resistance and, in the presence of nickel and chromium, increases the tensile strength and yield strength. Molybdenum makes forging more difficult, improves hardenability, fatigue resistance and magnetic properties. It has a notable influence on weld properties. It is a carbide-forming element. In high speed steels, it increases toughness while maintaining hot hardness and cutting retention properties. It replaces tungsten for the formation of carbides, in a ratio of 1% molybdenum to 2% tungsten.

e) Silicon - Si - melting point 1,410°C. Increases the yield strength of steels. Element that stabilises ferrite and thus reduces the formation of carbides, helping to decompose cementite (Fe_3C) into ferrite. It impairs elongation, toughness, thermal conductivity and machinability, but increases resistance to atmospheric corrosion. A steel can only be considered a silicon steel when the silicon content is greater than 0.60 per cent. Silicon steels have a good tempering capacity because they have a low critical cooling rate. Cold-rolled steel sheets with low carbon and high silicon contents have higher magnetic permeabilities and are used in electric motors and transformers.

f) Nickel - Ni - The addition of nickel gives the steel greater hardening penetration, as it considerably reduces the critical cooling speed. When alloyed with chromium, it increases the toughness of the processed steel. In large quantities, together with chromium, it makes the steel resistant to corrosion and heat. It has a direct influence on making the grain finer.

Alloy steels are commonly referred to by the name of the element or elements that influence their characteristics, regardless of their content or levels in the composition:

a) Steel - Nickel;
b) Steel - Nickel - Chrome;
c) Steel - Manganese;
d) Steel - Molybdenum;
e) Steel - Chrome - Nickel - Molybdenum;

It can therefore be said that soil penetration tools are made up of alloys whose purpose is to provide hardness, ductility and resistance to abrasion. The SPF manufacturer must seek a balance between these three characteristics, which can be modified according to their use.

Some manufacturers usually make a choice based on three parameters: Hardness; resistance to wear and penetration. Ideally, the material should be able to withstand these three parameters as well as the varying soils that occur in all mining operations.

4.7 - WEAR-RESISTANT HARD COATINGS

To recover components affected by metal loss (wear) during the production cycle, hardfacing is applied by welding using various consumables, techniques and processes.

Metal alloys applied by welding to surfaces in order to protect against wear and consequently increase the useful life of parts and equipment have been used on a large scale in the consumer goods, mining and sugarcane industries (Leite and Marques, 2009).

For the application of hardfacing by welding, tubular wires have become an increasingly viable alternative due to their high productivity and weld quality, partly replacing the use of coated electrodes.

Hardfacing is the name given to the process of producing a hard, wear-resistant layer on the surface of a part subject to wear by welding. The process can be applied to new parts as well as to the recovery of parts worn out by use. It is a simple process to apply, requiring only the coating alloy rods and an oxyacetylene flame or an electric arc. Hardfacing can be applied to most non-ferrous metals. On the other hand, with a few exceptions, hardfacing non-ferrous alloys with a melting point below 1100°C is not recommended (Baptista and Nascimento, 2014).

Carbon steels can be coated relatively easily, especially if the carbon content is less than 0.35 per cent.

Welding becomes more difficult as the carbon content increases, and high carbon and alloy steels must be pre-heated and annealed after coating.

Stainless steels, cast irons and high-speed steels can also be coated using appropriate welding techniques. Steels can be coated by welding, while copper, brass and bronze present greater difficulties due to their low melting points and high conductivity.

4.8 - CLASSIFICATION OF HARD COATING MATERIALS

According to Baptista and Nascimento in the Metals Handbook, 1964, hardfacing materials are classified into five large groups (Table 4.5), depending on the total content of alloying elements.

In general, wear resistance and cost increase with the group number. The choice of a specific alloy will depend on the application and the type of welding used. Group 1A alloys are low-alloy steels, almost always with chromium as the main alloying element. The total content of alloying elements, including carbon, is between 2% and 6%. These alloys are often used as a base layer for hardfacing with higher alloy content alloys.

The ferrous alloys in sub-group 1B are similar to those in group 1A, but with a higher content of alloying elements (6% to 12%) and a carbon content of up to 2% or more. Many tool steels and some alloys are part of this group.

Group 1 materials have the highest toughness among hardfacing alloys, particularly austenitic manganese steels. They also have better wear resistance than low and medium carbon steels, which are the alloys usually coated in group 1.

They are cheaper than other hardfacing alloys and are widely used where machinability and only a moderate increase in wear resistance are required.

Subgroup 2A alloys are alloys in which the main alloying element is chromium, with a total alloying element content of between 12% and 15%. Some alloys have a small percentage of molybdenum. This group also includes some non-ferrous castings with a medium alloy content. Molybdenum is the main alloying element in almost all materials in group 2B, which also have relatively high chromium contents.

Subgroup 2C is made up of austenitic manganese steels. Manganese is the main alloying element, while nickel is also present to further stabilise the austenite.

The alloys in subgroups 2A and 2B are more wear-resistant, less tough and more expensive than the materials in group 1. Subgroups 2C and 2D, on the other hand, are very tough but have limited wear resistance, which can be greatly increased by hardening. Subgroup 2D alloys have a total alloy element content of between 30% and 37%, with carbon varying between 0.10% and 1%.
Group 3 materials contain 25% to 50% alloying elements. These are high-chromium alloys, many of which contain nickel and/or molybdenum. The carbon content ranges from 1.75% to 5%. This group is characterised by the massive presence of hypereutectic carbides, which give the alloys high wear resistance and reasonable resistance to corrosion and heat. These alloys are more expensive than those in groups 1 and 2.

Table 4.5 - Classification of hard coating materials.

	Total content of alloying elements %	**Main Alloying Elements**
	Low Alloy Ferrous Materials	
1 A	2a6	Cr, Mo, Mn
1 B	6a 12	Cr, Mo, Mn
	High Alloy Ferrous Materials	
2A	12a 15	Cr, Mo
2 B	12 a 25	Mo, Cr

2C	12 a 25	Mn, Ni
2 D	30 a 37	Mn, Cr, Ni
3A	25 a 50	Cr, Ni, Mo
3 B	25 a 50	Cr, Mo
3C	25 a 50	Co, Cr
	Nickel- and cobalt-based alloys	
4A	50 a 100	Co, Cr, W
4 B	50 a 100	Ni, Cr, Mo
4C	50 a 100	Cr, Ni, Mo
	Carbides	
5	75 a 96	WC or WC in combination with other carbides.

Source: Metals Handbook, v.2, 1964

Group 4 comprises cobalt-based and nickel-based alloys with a total non-ferrous element content of between 50% and 99%. Cobalt-based alloys, subgroup 4A, are considered to be the most versatile hardfacing materials. They resist heat, abrasion, corrosion, impact, flaking, oxidation, thermal shock, erosion and metal-to-metal wear.

Some of these alloys maintain high hardness up to 825°C and resist oxidation up to temperatures of around 1100°C. Nickel-based alloys, subgroup 4B, are particularly suitable for parts subject to both corrosion and wear. They are superior to other hardfacing materials in applications where wear is caused by metal-to-metal contact, such as bearings. They maintain high hardnesses up to around 650°C and resist oxidation up to temperatures of around 875°C.

Group 5 materials consist of hard granules of carbides distributed in a metallic matrix. They are very suitable for intense abrasion and cutting applications. Initially, only tungsten carbides were used. More recently, other carbides, mainly titanium, tantalum and chrome, have been used with good results.

Various materials have been used as a metal matrix, such as iron, carbon steel, nickel-based alloys, cobalt-based alloys and bronzes. Group 5 materials have the highest resistance to abrasion in parts subjected to small or moderate impacts.

4.9 - MATERIAL SELECTION FOR HARD COATING

The first step is to thoroughly characterise the service conditions in which the hardfacing will have to work. Then carefully analyse the interactions between the candidate materials and the base metal, as well as the welding process to be used for the hardfacing. A general guide for selecting hardfacing alloys, prepared by the Metals Handbook, is reproduced in Table 4.6.

For proper selection, the following points must be observed (Baptista and Nascimento, 2014):

a) Analyse the service conditions to determine the types of wear resistance and environmental resistance required;
b) Select some candidate materials;

c) Analyse the compatibility between these alloys and the base metal, including considerations of thermal stresses and possible cracking;
d) Test components coated with the candidate materials;
e) Selecting the best alloy, taking into account both the cost and duration of the coating.

Table 4.6 - Guide to selecting alloys for hardfacing.

Conditions of Service	Hard Coating Materials
Metal-to-metal slippage; high contact stresses	Stellitel, Tribaloy alloys
Metal-to-metal sliding; low contact stresses	Low-alloy steels for hardfacing
Metal-to-metal slippage combined with corrosion and oxidation	Cobalt or nickel-based alloys, depending on the aggressiveness of the environment
Abrasion under low stress; erosion by collision of particles at a small angle	High alloy cast irons
Severe abrasion under low stress, edge retention	Materials with high carbide content
Cavitation and collision erosion	Cobalt-based alloys
Intense mechanical shocks	High-alloy manganese steels
Intense mechanical shocks combined with corrosion or oxidation	Stellite 21, Stellite 6
Groove abrasion - Peeling	Austenitic manganese steels Stellite 21, Stellite 6, Tribaloy T-400, Tribaloy T-800
Thermal stability and/or creep resistance at high temperatures	Cobalt-based alloys, nickel alloys with carbides

Source: Metals Handbook, v.2,1964

Select the hardfacing process, taking into account the deposition rate, degree of dilution (the base metal dilutes the hardfacing alloy during welding), deposition efficiency and total cost. The total cost includes the price of the electrodes and processing.

4.10 - TYPES OF COATING

Coating through the welding process is a technique used for surface protection or to increase the material's resistance to abrasive wear when equipment is under severe operating conditions. The process consists of depositing materials on parts and components that are resistant to the stresses to which the material will be subjected.

Dilution ensures that the coating retains the properties of the alloy to be deposited and prolongs the useful life of the part or component. The amount of filler material deposited must be sufficient to guarantee a metallurgical bond.

The types of coating most commonly used in welding can be categorised as follows:

a) *Hardfacing* This process is the deposition of a hard, wear-resistant material on the surface of a less noble material (substrate) through a welding process. Its main function is to reduce wear, i.e. the loss of material due to abrasion, impact, erosion, cavitation or any other deterioration due to friction (Davis 1993).

   Surface hardening is also widely used to control the combination of wear and corrosion (Davis 1993). Another factor is when an alloy is homogeneously deposited by welding on the surface of a steel.
   soft material (low or medium carbon steel) in order to increase its hardness and wear resistance without causing a significant loss of toughness or ductility in the substrate. It is applied to reduce wear, abrasion, impact, erosion, galling or cavitation.

   According to Hutchings (1992), hardfacing is an alloy homogeneously deposited by welding on the surface of a soft material, usually a low or medium carbon steel, with the purpose of increasing its hardness and resistance to wear, without causing significant loss of ductility and toughness of the substrate.

   Conde (1986) states that hard coatings are used to reduce wear due to abrasion, erosion, impact or cavitation.

b) *Build up: This* involves adding metal by welding in order to restore the original dimensions of the component. In this case, the strength of the weld is the most important prerequisite of the project. It refers to the addition of metal by welding on the surface of the base metal in order to restore the original dimensions of the component. Weld strength is an important prerequisite that must be considered in the design (Conde, 1986).

c) *Cladding* consists of applying a corrosion-resistant metal to another metal whose corrosion resistance is inferior or was not taken into account in the design. They are generally used to deposit a sheet of filler metal on a low alloy carbon steel in order to provide surface protection against environmental corrosion, when in general the resistance of the coating is not included in the design of the component. This type of coating can be applied to resist localised corrosion ("pitting"), crevice corrosion, intergranular corrosion and low stress corrosion (Lima, Aldemi, 2008).

The corrosion resistance of the coating, in many cases, is the limiting factor in the life of the component, and is therefore the first consideration to be taken into account when selecting the alloy, the welding process and the procedure to be used (Conde, 1986).

d) *Buttering:* A coating method that provides a high ductility layer of solder before the actual welding. One or more passes of solder are applied to the face of the joint. The difference from the previous case is that it is used for metallurgical reasons and not for dimensional control. A classic example is the deposition of high-nickel alloys on a low-alloy steel base. The buttered part can be used in the as-welded condition or undergo subsequent heat treatment to obtain special properties (Conde, 1986).

CHAPTER 5

METHODOLOGY

In accordance with the objective proposed for this research project, which involves analysing the development of large excavator tips, with and without hard material coating, the following general aspects were taken into consideration:

a) Data collection on the mining company's existing excavators and loaders, which work with ground-penetrating tips, in order to target the most cost-effective tip.
b) Evaluation of the tip shapes available on the mining market and definition of the tip with the best performance for working in mine soils;
c) Definition of SPF materials developed on the market for the soil characteristics under study, altering steel alloys to improve impact and wear resistance;
d) Hardfacing using the coated electrode welding process with the aim of increasing the hardness and wear resistance of the tips.

5.1 - INITIAL DATA COLLECTION

Based on the information generated by the software database that the company uses to control the parts in the equipment, an analysis was made of the tip sets that had the highest cost over the period of one year of operation. At this initial stage, the useful life of each tip was not assessed.

The software the company uses to generate this data is commercial software called SAP Software. Using the data from this software, it was possible to check the values, quantities and hours worked by each set of tips.

Figure 5.1 shows the sequence of tests in the experimental flowchart.

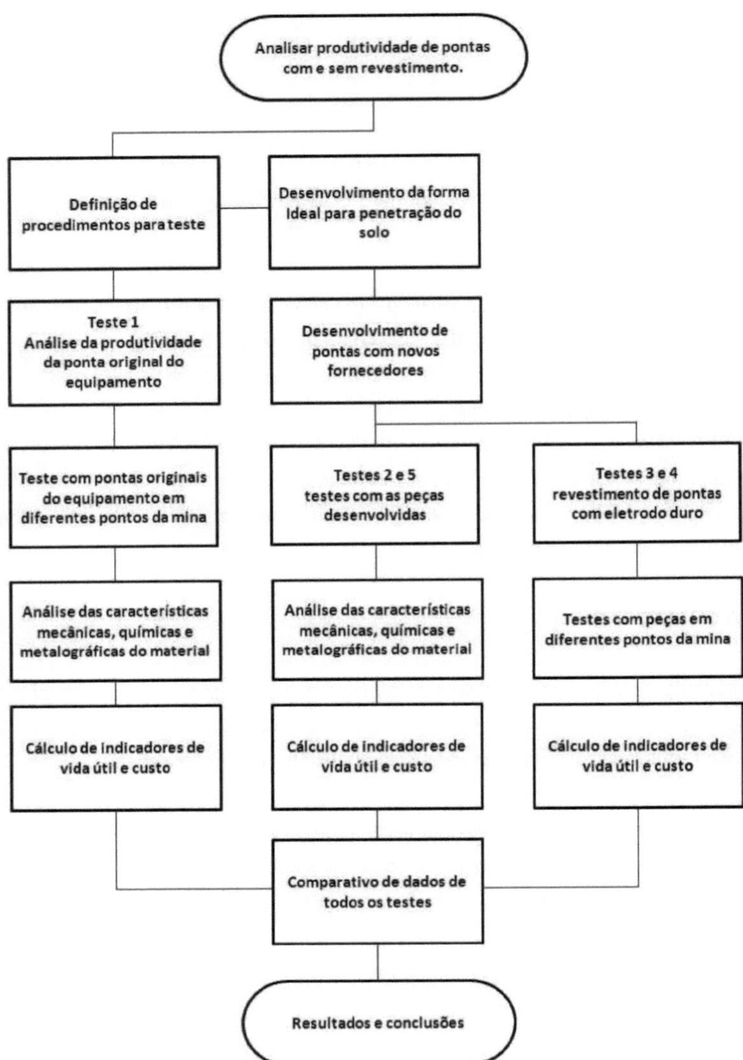

Figure 5.1 - Experimental flowchart
Source: Author

5.2 - DEFINING TEST PROCEDURES

According to the initial data collection, Table 5.1 shows the fleets of equipment chosen as the object of study. These are excavators (code ESC) and loaders (code CR) that are used in the mines in various operations and soils. They have the following characteristics:

a) Defined by the models and their respective weights in tonnes
b) Composed of 2 to 5 pieces of equipment per fleet.
c) The equipment as a whole has between 04 and 08 tips, as shown in Table 5.1.

Table 5.1 - Mining equipment containing the SPFs in the study.

Model	Quantity	Group	Tips per set
25 tonne wheel loader	5	CR	08
50 tonne wheel loader	5	CR	08
78 tonne wheel loader	2	CR	08
41 tonne excavator	3	ESC	05
75 tonne excavator	2	ESC	04
80 tonne excavator	3	ESC	04
110 tonne excavator	4	ESC	04

Source: Author

Once the data has been collated, the equipment with the highest cost to the company in one year of operation is chosen for the study. The cost of the items is related to all the equipment in the fleet. Once the fleet with the highest cost has been defined, the productivity tests begin.

The equipment purchased by the mining company works with a set of original tips that often don't match the soil being worked. The tests were carried out on the same equipment in the fleet. The only difference in the tests carried out was in the soils of the mine, given the diversity encountered. After defining the study equipment, some test procedures were defined:

a) Choose one of the machines in the fleet to test for all tests;
b) Installation of a set of tips collecting the initial hour meter;
c) Normal production operation at various locations until replacement is required;
d) Monitoring the length and width of the ends until they reach the point where they can be removed from the machine;
e) When exchanging the set for a new one, collect the current hour meter to calculate the useful life and cost of the previous worn-out set. The initial hour meter is used for the next set installed;
f) Due to the variety of soils, each test was carried out with 11 sets of tips, which lasts an average of three months of operation;
g) Sending samples of tips, one new and one worn, for analysis;
h) Analysis of the mechanical characteristics of the new tip using tensile, hardness and initial weight tests;
i) Analysis of the chemical characteristics of the new tip, providing information on the elements

present in the sample;

j) Analysing the micro-graphic characteristics of the new tip to assess the structure of the steel;
k) Comparison of useful life and cost per hour worked indicators highlighted in section 5.2.1;
l) Comparison of test characteristics.

It's important to note that the hour meter is the instrument installed in the equipment to inform you of the equipment's operating hours. It only starts counting when the equipment is "on", i.e. as soon as the diesel engine is started. When the equipment is switched off, the hour meter stops counting.

The five tests carried out on the tip assemblies are:

1) Test with the equipment's original stylus set;
2) First test to develop a set of new tips by modifying the shape and material;
3) First test with hard material coating with bead spacing carried out on the tip developed in item "2" above;
4) Second test with hard material coating without spacing between the strands carried out on the tip developed in item "2" above;
5) Second test to develop a set of new tips by modifying the shape and material.

The full testing procedure was followed in three of them (tests 1, 2 and 5). In these three tests we have material/shape development, the first being original equipment. The other two tests (3 and 4) are just surface coating with a hard material electrode and do not require steps "h", "i" and "j" of the procedure, since the coating is done on a developed tip.

1.1.1 - INDICATORS FOR ANALYSING USEFUL LIFE AND COST

The work involved carrying out mathematical calculations in order to define the best cost/benefit of the tips chosen and the useful life of the set. These were carried out using the equations below. These formulas are constantly used in mining operations to evaluate these indicators, since, as previously stated, they are among the biggest costs in the industry.

$$\mathrm{VU} = \frac{\mathrm{HT}}{\mathrm{Q}}$$

$$\mathrm{CT} = \frac{\mathrm{R\$}}{\mathrm{HT}}$$

Where:

VU = Useful life of the tips (hours);

CT = Cost of the set per hour worked on the equipment (reals per hours worked);

HT = hours worked of the tips by the sets;

Q = Quantity of tip sets (unit);

R$ = Average cost realised at the excavator fleet ends (reals).

5.3 - TEST 1 - ORIGINAL TIP

5.3.1 - ANALYSING THE PRODUCTIVITY OF THE ORIGINAL EQUIPMENT TIP

Once the equipment with the highest cost had been selected and a piece of equipment from the fleet had been chosen, tests began according to the procedure described in item 5.2 with the original set of tips that already worked with the equipment. After the field tests, two tips, one new and one worn, were analysed in the laboratory to collect the mechanical and chemical characteristics and micrographs of the material. The two mining indicators (as per item 5.2.1) were also calculated to check the cost-benefit ratio.

5.3.2 - TESTING WITH ORIGINAL EQUIPMENT TIPS AT DIFFERENT POINTS IN THE MINE.

Due to the wide variation in soil, there is a variation in the useful life of the tips, which is around 50 hours for more compacted, high-friction soils and 300 hours for friable (loose), low-friction soils. Figures 5.2 and 5.3 show the differences between the two soils that the excavators and loaders work on, i.e. more compacted, high-friction soils and friable (loose), low-friction soils respectively.

Figure 5.2 - Excavator working in more compacted soils with high friction.
Source: Mineração Usiminas - Mina Oeste - Itatiaiuçu - 2013

Figure 5.3 - Excavator working on friable (loose) soils with low friction.
Source: Mineração Usiminas - Mina Oeste - Itatiaiuçu - 2013

As shown in Figure 5.2, the soil is very compacted, with several rocks, causing greater impacts and abrasion, wearing out the assembly which can last only 50 hours. In Figure 5.3, the soil is completely loose, reducing impacts and abrasion and increasing the useful life to up to 300 hours.

5.3.3 - ANALYSIS OF THE MATERIAL'S MECHANICAL, CHEMICAL AND METALLOGRAPHIC CHARACTERISTICS.

Below are Tables 5.2, 5.3 and 5.4, which were filled in during the tests to provide information on the mechanical, chemical and metallographic characteristics of the tip assemblies tested, respectively. After collecting the information from tests 1, 2 and 5, the values were compared in a single table to evaluate the results.

Table 5.2 - Table with the material's mechanical characteristics.

Description	Value
Tensile Strength (MPa)	
Flow Flow (MPa)	
Elongation (%)	
Stricture (%)	
Initial Hardness (HB)	
Final Hardness (HB)	
Initial weight (Kg)	

Source: SENAI - Itaúna CETEF Marcelino Corradi - 2013

Table 5.3 - Chemical composition of the material in % by weight.

Description	Value
C - Carbon (%)	

Si - Silicon (%)	
Mn-Manganese (%)	
Cr - Chromium (%)	
Mo - Molybdenum (%)	
Ni - Nickel (%)	

Source: SENAI - Itaúna CETEF Marcelino Corradi - 2013

Table 5.4 - Micrographic analysis of the SPF.

Description	Features
Type of steel	
Heat treatment	
Micro structure	
Observation	

Source: SENAI - Itaúna CETEF Marcelino Corradi - 2013

5.3.4 - CALCULATING USEFUL LIFE AND COST INDICATORS.

The indicators were entered into Table 5.5 so that they could later be compared with the other materials.

Table 5.5 - Cost and useful life of tips.

Tip	Number of sets	Cost (R$ / HT)	Service life (HT / Q)
Original tip			

Source: Author

5.4 . - DEVELOPING THE IDEAL SHAPE FOR PENETRATING THE GROUND

As mentioned earlier, there is a wide variety of tip shapes on the market. For mining soils, there are three tips that stand out because they have good wear resistance and good impact resistance, which are far superior to the others.

Figure 5.4 shows the reinforced penetration tip. This tip has the advantage of excellent soil penetration, as well as good mechanical strength and good impact resistance. Although it has these advantages, this tip does not have a long service life compared to other tips.

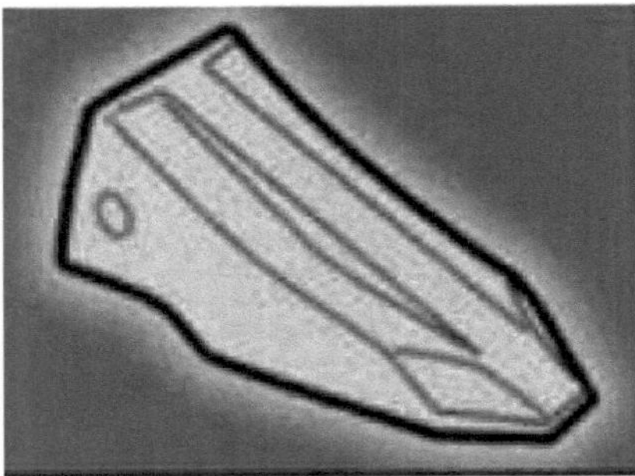

Figura 5.4 - Reinforced penetration tip.

Source: *http://carajasmaxxi.com.br/pecas fps pontas.htm*, **2014**

Figure 5.5 shows the reinforced long-life tip. This tip has the advantage of good mechanical strength and good impact resistance. Its disadvantage is poor penetration compared to the other tips.

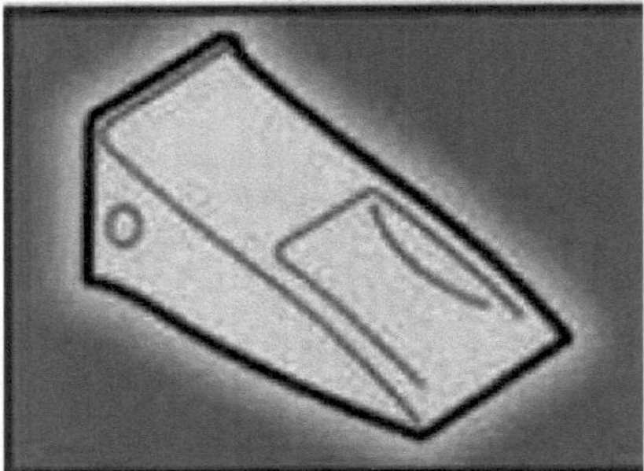

Figura 5.5 - Reinforced long-life tip.

Source: *http://carajasmaxxi.com.br/pecas fps pontas.ht, 2014*

Figure 5.6 shows the long penetration life tip. This tip has the advantage of good penetration, good mechanical strength and good impact resistance. Despite this good penetration, being pointed and having higher productivity, its disadvantage is the need to rework the tip, i.e. inserting hard material into the centre of the tip, thus increasing costs.

Figura 5.6 - Long penetration life tip.

Source: *http://carajasmaxxi.com.br/pecas fps pontas.htm, 2014*

Another factor that needs to be analysed in the tips is their mounting, as they can be pin-mounted or plug-mounted. This analysis is necessary because it is from this that the ease of assembly and the need to develop other tips will be verified.

5.5 - DEVELOPING LEADS WITH NEW SUPPLIERS

To develop the new tip for this study, a detailed design was carried out, as well as metallographic analysis of all the materials used to manufacture it. In this study, the shape of the original tip is analysed and the shape of the tips chosen for the study is compared.

5.6 - TESTS 2 AND 5 - DEVELOPED TIPS

The analyses of tests 2 and 5 follow the same sequence as test 1, going through the topics reported in items 5.3.1, 5.3.2 and 5.3.3.

The second supplier was chosen because it has developed a new hard material technology, specifically applied to ground penetrating tools (FPS). Combined with the new tip shape (reinforced long life), it showed excellent results. In order to achieve the desired wear control and impact resistance, the use of combinations of various alloying elements and heat treatments pertinent to each metal's chemical composition was necessary.

The indicator calculations for tests 2 and 5 are the same as for test 1, as per item 5.3.4.

5.7 - TESTS 3 AND 4 - COATINGS

The tip coating process is called "tip chiselling".

The tip chiselling process was carried out using electrodes with a harder material than the hardness of the tip. The way in which the tip is "chiselled" must be taken into account, as it can influence the tip's service life.

To increase the surface resistance of the new tip and consequently increase its useful life, it was hard coated using the electric arc welding process. The filler material used was a coated electrode with a hardness greater than the hardness of the tip. This coating electrode was deposited on the base metal, forming a new surface with a layer of approximately 5.0 mm. The hardness of the filler material is 60 to 62 HRc, i.e. around 654 to 688 HB.

The spacing between the strands in this ploughing is an important parameter, as it will guarantee greater or lesser performance from the tip.

In this work, test 3 was carried out with tips with a 40 mm bead spacing and test 4 with a bead spacing without bead spacing. The indicator calculations for tests 3 and 4 are the same as for test 1, as described in 5.3.4.

AWS E FeCr-A1 electrode with a hardness of 58 HRc was used to carry out the first chiselling test on the excavator tip. This electrode was chosen because its application is specifically for coating the teeth and edges of excavator buckets, gum blades, mill hammers and so on.

Other important characteristics of this consumable, which confirmed that it was the best option for use in this work, are that it is a rutile electrode with a high chromium carbide content, it has excellent weldability, an excellent finish and it also has

excellent resistance to abrasion and erosion wear, combined with moderate impacts.

In the second coating test for the deposition sequence of the hard coating layers, a high-chromium cast iron electrode was used. This type of electrode is used for components subject to severe and heavy abrasion and impact. The deposit made with this electrode features finely dispersed titanium carbide particles in a high chrome martensitic matrix.

The following procedure was used to carry out the tests on the tip with checker 2. Two types of tip were placed on the same excavator: on one side, two original tips were left and on the other side, two tips prepared with checker 2 were placed (Figure 6.7). It was also decided that the tips analysed should be the two central tips of the bucket, as the tips at the ends suffer different stresses because the operators of the excavator often use these tips to remove rocks and blocks.

This procedure was adopted in order to ensure that the two tines were being subjected to the same working conditions as well as excavating the same soil.

5.8 - COMPARISON OF DATA FROM ALL TESTS.

Once the tests were completed, comparative tables of mechanical, chemical and metallographic characteristics were made to compare tests 1, 2 and 5, as these are tests to assess the material of the tip assemblies.

Another comparison involves all five (5) tests. The useful life and cost indicators of all of them are compared to find the test with the best cost-benefit ratio, i.e. the test with the best useful life and the best cost per hour worked.

5.9 - TESTS USED TO EVALUATE THE CHARACTERISTICS OF THE TIPS

After testing the equipment, the tips were sent to a company specialising in laboratory analysis to describe their mechanical and chemical characteristics:

5.9.1 - MECHANICAL SURFACE HARDNESS TEST

This test was carried out on the new part and the worn part.

a) Unit of Measurement: Brinell Hardness (HB);
b) Load Used: 187.5 kg
c) Penetrator: 2.5
d) Durometer Used: Universal Dura Vision DV30, calibrated by a laboratory belonging to the Brazilian Calibration Network - RBC
e) Test carried out in accordance with Standard NBR NM ISO 6506-1:2010
f) Ambient temperature in the laboratory: around 22.0 °C

5.9.2 - MECHANICAL TENSILE STRENGTH TEST

This test was only carried out on the new tips.

a) Tensile strength in MPa;
b) Flow resistance in MPa;
c) Elongation of the material during the test;
d) Strictness of the material during the test
e) Initial specimen length = 50 mm
g) Force Generator Universal Traction Testing Machine, calibrated by a laboratory belonging to the Brazilian Calibration Network - RBC
h) Test carried out in accordance with ABNT NBR ISO 6892-1:2013 B:
i) Laboratory ambient temperature around 22.0 °C.

5.9.3 - CHEMICAL ANALYSIS BY OPTICAL SPECTROMETRY

This analysis aims to provide information on the chemical composition of the tip materials, recording the concentration of each element, such as: Carbon (C), Silicon (Si), Manganese (Mn), Phosphorus (P), among others.

a) Laboratory ambient temperature around 22.0 °C;
b) Analysis carried out in accordance with technical instruction IT Lab 249, revision 04;
c) Method used - Optical Emission Spectrometry;
d) SPECTROMAX optical emission spectrometer, calibrated by the manufacturer

CHAPTER 6

RESULTS AND DISCUSSIONS

6.1 - INITIAL DATA COLLECTION

Table 6.1 shows the percentage cost of each tip by size of equipment in relation to the total spent on ground penetrating tips alone, on the equipment, in one year of operation. This data was obtained from the SAP software used by the company.

Table 6.1 - Percentage of tip costs distributed by type and size of equipment. Data collected over one year of equipment operation.

Model	Quantity	Group	% Costs
25 tonne wheel loader	5	CR	7
50 tonne wheel loader	5	CR	8
78 tonne wheel loader	2	CR	11
41 tonne excavator	3	ESC	2
75 tonne excavator	2	ESC	9
80 tonne excavator	3	ESC	28
110 tonne excavator	4	ESC	35
Total			100%

Source: SAP Software - Usiminas Mining, 2013

Based on the data collected, it was found that the large 110 tonne excavator in the ESC group had the highest cost, so this was the equipment chosen as the object of study for this work. All the equipment in this group (ESC 110 tonne) operates in all parts of the mine, in other words, it works in all types of soil.

The equipment in this fleet had been in operation for less than a year, yet the cost was higher than the other equipment.

6.2 - EXECUTION OF THE TEST PROCEDURE

The equipment started operating with the original set of tips and all the procedures outlined in item 5.2 were followed, including item "g", which asks for tips from this set to be sent for analysis. A new, unused original stylus was sent and another stylus from the same manufacturer, already worn, so that their material could be characterised. Figure 6.1 shows the two tips.

Figure 6.1 - New tip (larger) and worn tip (smaller).

Source: Author

It is important to note that the worn tip has retained the still pointed shape of the original tip, especially in the front width of the tip. This shape increases the tip's productivity, as it still has the power to penetrate and start up slopes, but it wears down very quickly, as the contact area is larger and the effort, which causes greater abrasion, wears down the whole faster than a bulging tip.

6.3 - TEST 1 - ORIGINAL TIP

6.3.1 - ANALYSING THE PRODUCTIVITY OF THE ORIGINAL EQUIPMENT TIP

We monitored the testing of the 11 sets of equipment at different points in the mine, passing through compact soils and friable soils. These tests took an average of three months. All the steps in the procedure described in item 5.2 were followed.

After production, two tips were separated, one new and one worn, so that the analyses could be carried out.

6.3.2 - ANALYSIS OF THE MATERIAL'S MECHANICAL, CHEMICAL AND METALLOGRAPHIC CHARACTERISTICS

The analyses were carried out in accordance with the technical standards for each test. Table 6.2 shows the average results of the mechanical property analyses carried out on the original tip in

Figure 6.1, i.e. the new tip. The first tests were tensile and then the Brinell hardness test.

After analysis, it was found that the initial hardness when the tip was new was 499 HB, while the final hardness, i.e. when the tip was already worn, was 480 HB.

Tables 6.3 and 6.4 show the chemical composition and micrographic analysis respectively found in the original tip.

Table 6.2 - Average results of the mechanical property analyses carried out on the original tip.

Description	Value
Tensile Strength (MPa)	1175,2
Flow resistance (MPa)	1072,2
Elongation (%)	1,6
Stricture (%)	6,3
Initial Hardness (HB)	499
Final Hardness (HB)	480
Initial weight (Kg)	67

Source: SENAI - Itaúna CETEF Marcelino Corradi - 2013

Table 6.3 - Chemical composition of the material in percentage by weight.

Description	Value
C - Carbon (%)	0,28
Si - Silicon (%)	1,95
Mn - Manganese (%)	1,55
Cr - Chromium (%)	0,64
Mo - Molybdenum (%)	0,41
Ni - Nickel (%)	0,16

Source: SENAI - Itaúna CETEF Marcelino Corradi - 2013

Table 6.4-Micrographic **analysis of** the original tip.

Description	Features
Type of steel	Alloy steel, typically standardised
Heat treatment	Tempered and quenched
Microstructure	Typical of martensite, tempered and bainite
Observation	Wear resistance is obtained by the higher hardness not counting the hardening capacity of the austenite

Source: SENAI - Itaúna CETEF Marcelino Corradi - 2013

6.3.3 - CALCULATION OF USEFUL LIFE AND COST INDICATORS

Table 6.5 shows the cost and useful life of the equipment's original set of tips. These figures relate to one quarter of use, in which 11 sets of tips were used. The tests were carried out on a piece of equipment from the 110 tonne fleet and worked on various soils in the mine.

Table 6.5 - Cost and useful life of the original equipment tips.

Tip	Number of sets	Cost (R$ / HT)	Service life (HT / Q)
Original tip	11	41	92

Source: Author

The cost per hour worked (R$/HT) started at around R$41.00. The tip sets in question needed to be changed every 92 hours (on average). This useful life value is well below the proposed objective of changing the tips only during preventive maintenance, i.e. after approximately 250 hours of operation.

6.4 - DEVELOPING THE IDEAL SHAPE FOR PENETRATING THE GROUND

In partnership with a supplier (which we will now call supplier 1), a project was developed to build a new tip with the shape and dimensions that would improve the process. For this study, the first factor to be chosen was the shape of the tips.

6.4.1 - CHOICE OF TIP SHAPE

The shapes chosen for this work were those that perform best in mining. Figure 6.2 shows the shapes that were initially chosen with their respective technical characteristics.

The long life penetration tip, presented in the literature review, was discarded because it was much more expensive than the other two in the first selection.

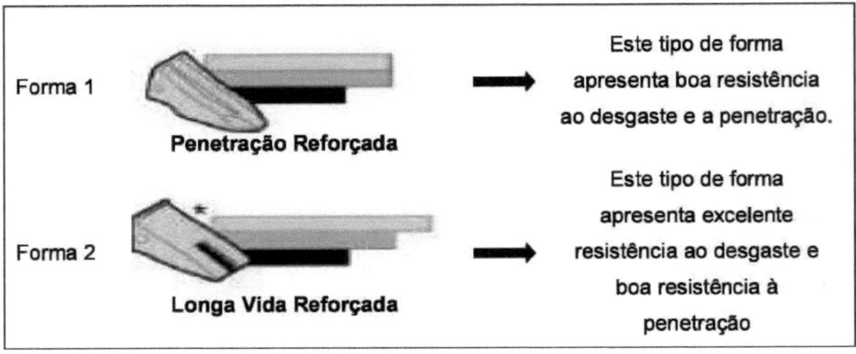

Figure 6.2 - Tips chosen for the initial tests.

Source: Escosoldering (2014)

After the initial analyses, it was decided that the reinforced long-life tip (Shape 2 in Figure 6.2) was chosen as the tip for the comparative analysis in this work.

In addition to the comparison, the informative colour graphs in Figure 6.2 next to each tip (tool wear bar graph) show that the wear resistance, highlighted in yellow (first bar), is higher in shape 2 than in shape 1.

Despite the less pointed shape, the grey bar graph (second bar) shows that the two tips have the same penetration and impact resistance (third bar).

6.4.2 - DEVELOPING TIPS WITH NEW SUPPLIERS

The first tip was dimensioned for testing and validation as shown in Figure 6.3. After approval, the test assemblies were made. One difference with this tip is that it has a more bulging shape than the original tip, as well as being heavier.

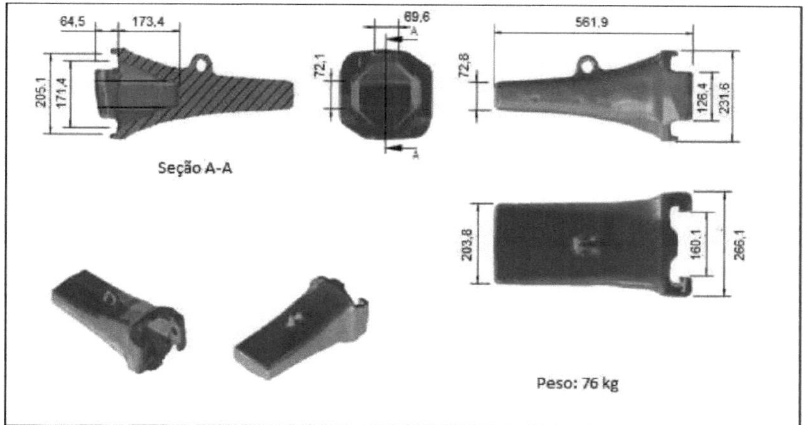

Figure 6.3 - Dimensions of the tip developed for the job.

Source: Supplier 1

An important detail of the dimensions chosen is the height of the tip nozzle, which is 72.8mm, and the width, which is 203.8mm. These dimensions were chosen because it is in this area that the tip has its first contact with the ground. The tip has a more robust front end, which reduces wear compared to a more pointed tip, but at the same time increases the effort the excavator needs to make to penetrate the ground. It's a greater effort, but within the machine's capacity.

6.5 - TEST 2 - SUPPLIER TIP 1

6.5.1 - MAKING TEST TIPS

The tip was manufactured according to the shape chosen and with the material indicated to meet the need, which was (ASTM A128 - Grade A steel). The main characteristic of this steel for the construction of this new tip was that it increases in hardness during the excavator's operating process. This increase in hardness is due to the formation of an austenitic microstructure with high impact resistance.

According to the data collected during the operation of this tip, it was found that this material starts the operation with a low hardness, i.e. 240HB, and finishes with a hardness much higher than the initial one, i.e. 500HB. This higher hardness at the end of the operation is due to the surface hardening process and the compressive stresses that occur during excavation.

Another important characteristic of this material is that it has a high percentage of manganese which contributes to increased hardenability, increased weldability as well as an increase in the tensile strength limit and a decrease in toughness. This set of properties achieved by increasing the percentage of manganese in the alloy leads to a considerable increase in abrasion resistance, which is the most important factor in increasing the tip's useful life. The weight of this tip was around 76kg, well above the original, mainly due to its square rather than pointed shape.

6.5.2 - TESTS WITH THE DEVELOPED TIPS

Right from the start of production with the reinforced long life tip (Shape 2), better performance was seen than with the reinforced penetration tip. On average there was an increase of over 20% in the hours worked, which defined the choice of this tip shape for the object of study. To obtain comparative values for the developed tip, a micro-structural analysis was made of the tip manufactured without use and a worn tip used in the test equipment. Figure 6.4 shows the two tips analysed, with the larger one in green being the new developed tip and the smaller one (grey) being the same tip already worn. Tables 6.6 and 6.7 show the results of the tensile tests and the chemical composition compared to the original tip.

Figure 6.4 - Newly developed and worn tip.

Source: Author

6.5.3 - ANALYSIS OF THE MATERIAL'S MECHANICAL, CHEMICAL AND METALLOGRAPHIC CHARACTERISTICS

Table 6.6 - Comparison of values between the original tip and that of supplier 1.

Description	Value Original tip	Value Supplier tip 1
Tensile Strength (MPa)	1175,2	860,9
Flow Flow (MPa)	1072,2	375,7
Elongation (%)	1,6	43,5
Stricture (%)	6,3	34,2
Initial Hardness (HB)	499	244
Final Hardness (HB)	480	483
Weight (Kg)	67	76

Source: SENAI - Itaúna CETEF Marcelino Corradi - 2013

Table 6.7 - Comparison of the chemical composition of the original material with the material from supplier 1 in percentage by weight.

Description	Value Original tip	Value Supplier tip 1
C - Carbon (%)	0,28	1,35
Si - Silicon (%)	1,95	1,00
Mn - Manganese (%)	1,55	14,00
Cr - Chromium (%)	0,64	1,00
Mo - Molybdenum (%)	0,41	0,00
Ni - Nickel (%)	0,16	0,00

Source: SENAI - Itaúna CETEF Marcelino Corradi - 2013

Table 6.8 shows the micrographic analysis of the original and supplier 1 tips.

Table 6.8 - Micrographic analysis of the original and supplier 1 tips.

Description	Features Original tip	Features Supplier tip 1
Type of steel	Alloyed steel, typically standardised.	Manganese steel type *HadField*
Treatment Thermal	Tempered and quenched.	Untreated.
Microstructure	Typical of tempered martensite and bainite.	Austenitic with low hardness.
Observation	Resistance to wear is obtained by the higher hardness not counting on the hardening capacity of the austenite.	High ductility, low flow and the CFC structure allow for high hardening capacity. This provides wear resistance for impact situations.

Source: SENAI - Itaúna CETEF Marcelino Corradi - 2013

6.5.4 - CALCULATION OF USEFUL LIFE AND COST INDICATORS

Table 6.9 shows the results of the costs and useful life obtained both for the tests carried out with the original tip that came with the new equipment and for the tests carried out with the tip developed with supplier 1. The tests lasted three months and 11 sets of tips were used.

Table 6.9 - Test results with original and developed tip supplied. 1.

Tip	Number of sets	Cost (R$ / HT)	Service life (HT / Q)
Original tip	11	41	92
Supplier 1	11	31	131

Source: Author

It can be seen that the original tip has a higher tensile and yield strength than the tip from supplier 1. Analysing these two parameters alone, the tip from supplier 1 would wear more with abrasion, but it can be seen that the elongation and stiffness are greater in the tip from supplier 1, which allows it to cushion itself before yielding or breaking.

Another important factor is hardness, which in the original tip decreases from 499HB in the new tip to 480HB in the worn tip. The opposite phenomenon occurs with the tip from supplier 1.

As the tip works, it increases in hardness. The weight is greater on the supplier 1 tip (76 versus 67 kg) due to the new rectangular shape.

Comparing the chemical composition, it was found that the main difference is in the percentage of manganese in supplier 1's tip, which is much higher than the percentage in the original tip.

According to the literature, the addition of manganese increases the hardenability, weldability and tensile strength, but decreases the toughness of the material. In large quantities it increases abrasion resistance.

According to the results shown in Table 6.9, it can be seen that there was a 26 per cent reduction in cost and a 42 per cent increase in service life when using the tip developed with supplier 1.

In view of these results, it is believed that the specification of the new tip fulfils the expectations set in terms of cost and partly in terms of the useful life of the tip. This is partly because the tip did not reach the 250 hours of useful life defined in the specific objectives.

6.6 - TEST 3 - RESULT / PERFORMANCE OF THE FIRST CHESS TEST

To make the weld seams, it was decided that they would be spaced 40 mm apart and deposited parallel to the penetration face, both on the underside and on the sides, as shown in Figure 6.5. This type of chequering has been referred to in this work as chequering 1.

In the upper part of the tip, above the eye, parallel strands were made next to each other. This deposition process was done manually.

Table 6.10 shows the results compiled for the three tests, i.e. testing the manufacturer's original tips, supplier 1's tip without coating and supplier 1's tip with hard material coating. Eight sets of tips were used for this test.

Figure 6.5 - Chiselling 1 of the tip made with the AWS E FeCr-A1 electrode.

Source: Author

Table 6.10 - Data compiled from the tips used for testing.

Tip	Number of sets	Cost (R$ / HT)	Service life (HT / Q)
Original tip	11	41	92
Tip Supplier 1 uncoated	11	31	131
Supplier tip 1 **with** coating 1	11	26 (obs)	140

Source: Author

It is important to note that the cost included the amount spent on inputs, as well as the labour costs required to coat the tips.

Analysing the results obtained in Table 6.10, it can be seen that there was an improvement in both the cost and the useful life of the tip that received the hard coating. However, the procedure for applying this hard coating requires more time to prepare the tip. In view of this, and the small variation in the results, it was concluded that it is a viable process, but that it requires more work to produce the tips.

It's worth remembering that the cost of the added material as well as the cost of labour is part of the final cost per hour worked. Despite reducing the cost per hour worked from 31 to 26 (R$ / HT) and increasing the useful life from 131 to 140 hours, it was decided to stop the tests after two months, but the 11 sets requested were used. Another factor that helped stop the test was the way the hard coating was applied, which was handmade.

The second checker test was carried out mechanically to ensure the standardisation of the beads, by coating the beads using an automatic welding machine.

6.7 - TEST 4 - RESULT / PERFORMANCE ACCORDING TO CHESS TEST

To try to improve the performance of the tip with the hard coating technique, a second chiselling test was carried out (chiselling 2) using a new deposition procedure, i.e. making several layers of coating. In this second test, it was decided to hard coat the entire front of the tip, as shown in Figure 6.6.

Figure 6.6 - Chiselling 2 of the tip made with the AWS E FeCr-A1 electrode.

Source: Author

The tips were coated by the electrode supplier and six sets of tips were prepared for the tests, all of which were used during the three months of testing.

Figura 6.7 - Excavator with two types of tip - original and ploughed 2.

Source: Author.

Table 6.11 shows the results compiled for the four tips, i.e. the manufacturer's original tips, supplier 1's tip **without** coating, supplier 1's tip **with** coating 1 and supplier 1's tip **with** coating 2.

Analysing the results obtained in Table 6.11, it can be seen that the results also showed a slight improvement in both the cost and the useful life of the tip that received 2. However, the procedure for applying this hard coating requires even more time than the time spent preparing the tip with 1, as well as a greater amount of deposited material. In view of this, and the small variation in the results, it was concluded that it is a viable process, but that it requires more work to produce the tips.

Table 6.11 - Data compiled from the tips used for testing.

Tip	Number of sets	Cost (R$ / HT)	Service life (HT / Q)
Original tip	11	41	92
Tip Supplier 1 uncoated	11	31	131
Supplier tip 1 **with** chess. 1	11	26 (obs)	140
Supplier tip 1 **with** chess. 2	11	22 (obs)	151

Source: Author

It is important to note that this cost includes the amount spent on inputs, as well as the labour costs needed to coat the tips.

6.8 - TEST 5 - SUPPLIER END 2

6.8.1 - NEW TEST TIPS

In partnership with a new supplier (which we will now call supplier 2), a project was developed to build a new tip with the shape and dimensions that would improve the process. The shape remains the reinforced long life. This new tip has the same design as the tip developed with the first supplier, but with some changes shown in Figure 6.8. The main changes are in the front width of the tip, which went from 203.8 to 211.7 mm, the front height of the tip, which went from 72.8 to 83.8 mm, and the weight, which went from 76 to 92 kg.

The specifications of the chemical elements and the temperature of the heat treatments were controlled so that a hardness of between 400 HB and 480 HB could be obtained.

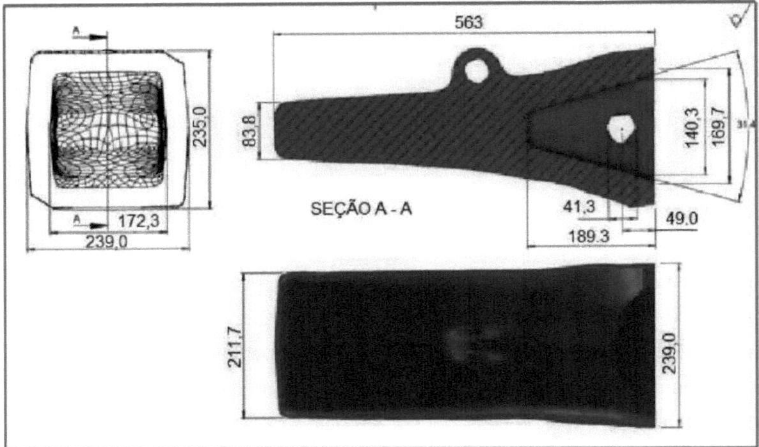

Figure 6.8 - Tips developed with supplier 2.
Source: Author

6.8.2 - ANALYSIS OF THE MATERIAL'S MECHANICAL, CHEMICAL AND METALLOGRAPHIC CHARACTERISTICS

Tables 6.12, 6.13 and 6.14 show the results obtained in the characteristics and mechanical tests, the chemical composition and micrographs of the original tip, the tip from supplier 1 and the tip from supplier 2.

Table 6.12 - Comparison of mechanical characteristics between the original tip, the tip from supplier 1 and the tip from supplier 2.

Description	Value Original tip	Value Supplier tip 1	Value Supplier tip 2
Tensile Strength (MPa)	1175,2	860,9	1429
Flow Flow (MPa)	1072,2	375,7	Nd

Elongation (%)	1,6	43,5	0
Stricture (%)	6,3	34,2	0
Initial Hardness (HB)	499	244	480
Final Hardness (HB)	480	483	477
Weight (Kg)	67	76	92

Source: SENAI - Itaúna CETEF Marcelino Corradi - 2013

Table 6.13 - Comparison of the chemical composition in percentage by weight of the original material with the material from supplier 1 and supplier 2.

Description	**Value** Original tip	**Value** Supplier tip 1	**Value** Supplier tip 2
C - Carbon (%)	0,28	1,35	0,21
Si - Silicon (%)	1,95	1,00	0,69
Mn - Manganese (%)	1,55	14,00	0,95
Cr - Chromium (%)	0,64	1,00	0,81
Mo - Molybdenum (%)	0,41	0,00	0,53
Ni - Nickel	0,16	0,00	1,55

Source: SENAI - Itaúna CETEF Marcelino Corradi - 2013

Table 6.14 - Micrographic analysis of the three tips tested.

Description	**Features** Original tip	**Features** Supplier tip 1	**Features** Supplier tip 2
Type of steel	Alloyed steel, typically standardised.	*HadField* manganese steel.	Alloy steel
Heat treatment	Tempered and quenched.	Untreated.	Hardened and tempered 02 times.
Structure	Typical of tempered martensite and bainite.	Austenitic with low hardness.	Martensite and bainite
Observation	Resistance to wear is obtained by the higher hardness not counting on the hardening capacity of the austenite.	High ductility, low flow and CFC structure allow for high capacity capacity hardening. This provides wear resistance for situations where situations involve impact.	Typical presence of inclusions at the ends of the sample and regions difficult to see by optical microscopy. It shows decarburization and inclusions typical of oxides and sulphides, and porosity typical of micro-ruptures scattered throughout the section examined.

Source: SENAI - Itaúna CETEF Marcelino Corradi - 2014

6.8.3 - CALCULATION OF USEFUL LIFE AND COST INDICATORS

Table 6.12 shows the material from supplier 2 with the highest tensile strength, no yield strength, elongation and stiffness. The hardness changed little from the new tip to the worn tip and the weight increased to 92kg.

There was a higher chromium and nickel content in the material from the second supplier, which gives the material greater tensile strength and tenacity, but you should be wary of impact breakage, as resistance to this phenomenon is lower.
Analysing the results obtained in Table 6.15, it can be seen that the tip from supplier 2 achieved a considerable improvement over the other four tips. It achieved a 66% reduction in cost and a 100% increase in useful life compared to the original tip that came with the equipment. Given these results, it is believed that this new manufacturing technology is viable for use.

The tips from supplier 2 showed greater material stability throughout the thickness, which allows for better penetration performance.

Table 6.15 - Data compiled from the tips used in the tests.

Tip	Number of sets	Cost (R$ / HT)	Service life (HT / Q)
Original tip	11	41	92
Tip Supplier 1 uncoated	11	31	131
Supplier tip 1 **with** chess. 1	11	26	140
Supplier tip 1 **with** chess. 2	11	22	151
Supplier 2 tip uncoated	11	14	184

Source: Author

6.9 - TIP EVALUATION

One test carried out after using the tips was the evaluation after removing the worn tips from the equipment. This is known as the Wear Test, which evaluates the variations in dimensions after removal to see if the set wears evenly.

A batch of tips was sent from supplier 1 after normal production. Figure 6.9 shows a set taken from the bucket and numbered according to the bucket assembly. The numbering follows the order from the first tip on the right side of the bucket, opposite the operator, to the fourth tip, which is the first tip next to the operator.

The percentage measured was the tip utilisation ratio, i.e. the new tip has a length of 561.9mm and has to be removed at a maximum of 180mm. The difference, 381.9mm, corresponds to 0% wear. As it wears down, it loses material until it reaches 180mm, which corresponds to 100% wear. The tip with less than 180mm can pierce and start to wear the adapter. As the assembly wears unevenly, i.e. the tips on the ends wear more quickly, and even on the same tip there is uneven wear on the front, analysing how the assembly is being removed allows you to see when it could still be produced with the assembly.

There is a big difference in the wear of the side tips compared to the centre tips. Figures 6.10, 6.11, 6.12 and 6.13 show the wear on each of them.

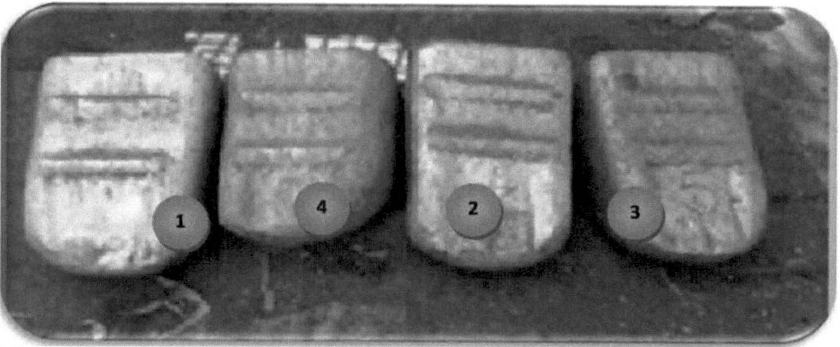

Figura 6.9 - Worn tips to be analysed.

Source: Author

Figura 6.10 - Tip 1 - Operator's side.

Source: Author

This tip showed 83% wear and has an irregular profile because it operates in one of the corners of the bucket (right side). It can be seen that one of the sides has worn down by more than 100% and has almost started to wear on the adapter.

Figura 6.11 - Tip 2 - Second tip in the sequence.

Source: Author

59.88% of this tip has been used up. This tip could still be utilised for more than 40 hours.

Figura 6.12 - Tip 3 - Third tip in the sequence.

Source: Author

72.07% of this tip has been used up. This tip could still be used for more than 20 hours.

Figura 6.13 - Tip 4 - Fourth tip in the sequence - tip on the operator's side.

Source: Author

This tip had an uneven wear profile because it operated in one of the corners of the bucket (left side). It wore 95% on one side and 103% on the other, very close to piercing and wearing out the adapter. Due to the wear on this tip, it was necessary to replace the assembly.

For best use, we recommend rotating the corner tips towards the centre.

The results obtained show that in the tests we obtained positive results and with each test carried out, the useful life was increased and the cost reduced. The tests involving hard electrode coating showed satisfactory results, but it should be borne in mind that more work is required to manufacture the tips.

CHAPTER 7

CONCLUSIONS

Based on the results obtained from the analyses of chemical composition, mechanical characteristics, micrographic analysis, useful life and cost, the following conclusions can be drawn:

The most suitable shape is the shape of the reinforced long life tip, as this tip showed greater productivity due to remaining longer with greater length and widths. It was the main change that improved the performance of soil penetration tools (FPS);

The cost/benefit ratio was achieved, as the cost was reduced from 41 to 14 reais per hour worked (R$/HT), which represents a 66 per cent reduction in cost;

J A considerable improvement was achieved in the useful life of the tip, as it increased by 100 per cent. The target was not reached, but the figures show satisfactory results and the prospect of achieving the expected result;

J It is necessary to rotate the tips on the sides of the bucket with the tips in the centre, as this allows the wear of the tips to be better controlled, preventing the bucket from being removed with the tips on the ends more worn than those in the centre. This rotation should take place halfway through the service life, i.e. currently around 90 to 100 hours of operation;

J With hard material coating on tips, an improvement in service life and a reduction in cost were obtained when applied to the tip of supplier 1, showing satisfactory results. No tests were carried out on the tips of supplier 2;

J The material with the highest tensile strength due to its higher chromium and nickel content performed better on the ground, but you have to watch out for impacts that can break the tips;

Due to the good results obtained on this type of equipment, this work should be multiplied to another excavator larger than 250 tonnes and can reach all the equipment in the excavator and loader fleets.

CHAPTER 8

SUGGESTIONS FOR FUTURE WORK

J Tests of changes in tip formats on all the equipment in the fleet;

J Tests with hard material coatings on the tips, partially coating the top of the tip so that the wear is on the side and the tip remains longer;

J Analysis of the shape of the tip in the diesel consumption of equipment;

BIBLIOGRAPHICAL REFERENCES

ABRAM, I.; ROCHA, A. V. Practical earthworks manual. Salvador/Ba. Sept. 2000.

ALBERTIN, E. Abrasive wear. 58th annual congress of the ABM - Brazilian Association of Metallurgy and Mining. Rio de Janeiro. 2003, p.55.

AWS. Gas Metal Arc Welding. Weilding Handbook & ed. Miami: AWS. v.2, 1991. P.109-1L55.

BAPTISTA, A. L. DE B., NASCIMENTO, I. A. Hard coatings resistant to wear deposited by welding used in the recovery of machine elements. Spectru Instrumental Científico Ltda. Rio de Janeiro. Website: www.spectru.com.br, 2014.

BARROS, M. B.; DE MELLO, J. D. B. Effect of test parameters on the predominant wear mechanism in abrasive wear tests. Revista Horizonte Científico, 2006, p.1-22.

BHOLE, S.D.; YU, H. Abrasive wear evaluation of tillage tool materials. Lubrication Engineering, v.48, n.12, 1992, p.925-34.

BUCHELY, M.F.; GUTIERREZ, J,C.; LEON, L.M.; TORO,A. The effect of microstructure on abrasive wear hardfacing alloys, Tribology and surfaces Group, National University of Colombia, MedelHn, Colombia. Science Direct. Wear 259.2005. p. 52-61.

CASQUET, Regis Quesada, PEREIRA, Walmir Carvalho. & LAGE, Edson Rogério. Mine Planning at Samarco Mineração S. A. Revista da Escola de Minas de Ouro Preto. 49 (3):set Ouro Preto: Escola de Minas, 1996. p. 7-10.

CASTRO, Cristóvão Américo Ferreira. Abrasive wear resistance of crawler tractor shoes after recovery processes. 73f. 2010. Dissertation (Master's Degree) Federal Technological University of Paraná - UTFPR, 2010.

Usiminas Heavy Plate CATALOGUE, 2014.

CONDE, R.H. Wear-resistant coatings. Boletín Técnico Conarco, n.85, 1986, p.2-20.

DAVIS, J.R. - Hardfacing, Weld Cladding and Dissimilar Metal Joinig. In: ASM Handbook - Welding, Brazing and Soldering, Vol. 6. 10th ed. OH: ASM Metals Park, 1993, p. 699-828.

DETTOGNI, Mareio Abbade. Main wear mechanisms and evaluation of different alloys for grinding bodies. 59 f. 2010. Monograph (Specialisation) Lato Sensu Postgraduate Diploma in Mineral Processing. Federal University of Ouro Preto School of Mines - Department of Mining Engineering, Ouro Preto, 2010.

DIN 50 320. Systematic analysis of wear processes: classification of wear phenomena. Metallurgy and Materials, v.53. 1997, p.619-622.

ESPÍRITO SANTO, A. C. Wear on the tips of no-till seed drill tines and its influence on tractive effort. Porto Alegre: UFRGS, Doctoral Thesis, 2005. p.152.

EYRE T. S. - Wear Characteristic of Metals, Source Book on Wear Control Technology, ASM, Metals Park, Ohio, 1978.

FERNANDES, J. C.; SANTOS, J. E. G.; SANTOS FILHO, A. G.; BORMIO, M. R. Evaluation of the wear of agricultural implements for different types of soils. In: Brazilian Congress of Agricultural Engineering, 31, Salvador. Proceedings... Salvador: SBEA, 2002. CD Rom.

GAHR, Karl-Heinz Zum. Microstructure and wear of materials. Tribology series, v10. Amsterdam: Elsevier, 1987. p.560.

GREGOLIN, J.A.R. Development of wear-resistant Fe-C-Cr-(Nb) alloys. Campinas : State University of Campinas, thesis (Doctorate) - Faculty of Engineering of Campinas/UNICAMP, 1990, p.228.

GUERRA, P. A. G. Operational Geostatistics. Brasília: DNPM, 1988. p.145.

GUIA DE COMPRAS - Brazilian Steel and Non-Ferrous Metals, 2014, p. 121;

HETTIARATCHI, D.R.P.; WITNEY, B.D.; REECE, A.R. The calculation of passive pressure in two-dimensional soil failure. Journal of Agricultural Engineering Research, v.11, 1966, p.89-107.

LIMA, Aldemi Coelho. Study of the application of hardfacing by welding with tubular wires in relation to the wear resistance of sugar cane chopping knives, thesis for a PhD in Mechanical Engineering, 2008.

http://pt.wikipédia.orq/wiki/Mineração in the open

http://pt.wikipedia.org/wi ki/Mining

http://www.asmaquinaspesadas.com/2012/Q6/fotos-e-videos-da-maior- excavator.html.

http://c3manuais.blogspotcom.br/2012/10/curso-escavadeira-hidraulica-case-cx.html

http://www.excavatorbucket.en.alibaba.eom/product/1763067166-221940381
/Kobelco SK350 Underground Mining Bucket for Excavator.html

http://carajasmaxxi.com.br/pecas fps tips.htm

http://www.ebah.com.br/content/ABAAABBn4AL/lavra-mina-ceu-aberto-subterranea

http://pt.dreamstime.com/imagem-de-stock-máquina-escavadora-que-enche-o- truck-image31565251

HUTCHINGS, I. M. Tribology: friction and wear of engineering materials, London: Edward Arnol, 1992.p.273.

LEITE, R. V. Melo and MARQUES, P. V. Comparative study of the resistance to abrasive wear of the coating of three metal alloys used in industry, applied by welding with tubular wires. *Soldag. insp. (Impr.),* vol.14, no.4, 2009, p.329-335.

MACHADO, Antônio L. T. et al. Methodology for evaluating wear on symmetrical tools for tillage. Revista Brasileira de Engenharia Agrícola e Ambiental, v.13, n.5, 2009, p.645-650.

MARTINS, F. A. Cladding welding with tubular wire. Master's dissertation. UFSCar. PPG. São Carlos. SP. 1995.

MESQUITA, R. A.; BARBOSA, C. A. An evaluation of wear and toughness properties in cold working steels. Tecnologia em Metalurgia e Materiais, v.2, n.2, 2005, p.12-18.

METALS HANDBOOK-v.2. Heat Treating Cleaning and Finishing. 8th Edition, Metal Park, Ohio, American Society for Metal, ASME, 1964.

MOURAD, R. B. A.; SANTOS, J. E. G. Design and construction of a bench to check the wear of the active parts of agricultural implements subjected to abrasion in four types of soil. Engenharia Agrícola, v.23 n.3, 2003, p.547-555.

NRM, Norma Reguladora de Mineração. Published by the National Department of Mineral Production, 2002.

NUNNALLY, S. W. Construction methods and management. Ed. Prentice Hall, New Jersey, USA. 2011.

OWSIAK, Z. Wear of symmetrical wedge-shaped tillage tools. Soil Tillage Research, v.50, 1997, p.295-308.

PASSOS, E.A.; LAGE, A.; PRADO, G. GTFPS. Technical group for ground penetrating tools. Vale do Rio Doce, copper department. Pará, 2010. p.61.

PEURIFOY, R. L.; SCHEXNAYDER, J. C.; SHAPIRA, A. AND SCHMITT, R. L. Construction, planning, equipment and methods. Ed. McGraw-Hill. New York. USA. 2011.

PINTO, T. P. Methodology for the differentiated management of solid urban construction waste. São Paulo. Thesis (doctorate) - Polytechnic School, University of São Paulo, 1999, p.189.

PINHEIRO, João Cesar de Freitas, A mineração brasileira de ferro e a reestruturação do setor siderúrgico. PhD Thesis State University of Campinas, 2000.

QUEIROZ FILHO, Cid Gomes, PEREIRA, Walrnir Carvalho & ALVES, Wagner Milagres. Productivity in mining. Belo Horizonte: IBRAM, 1997. p.10.

QUEVEDO, J. M. G.; DIALLO, M.; LUSTOSA, L. J. Simulation model for the loading and transport system in an open-cast mine. Pontifical Catholic University of Rio de Janeiro. Master's dissertation. Postgraduate Programme in Production Engineering. 2009. p.133.

Revista Manutenção & Tecnologia, Ed. 181, São Paulo, 2014, p.67.

Revista Manutenção & Tecnologia, Ed. 147, São Paulo, 2011, p.38.

RICARDO, H. S. and CATALANI, G. Practical Excavation Manual: Earthworks and Rock Excavation. Ed. PI NI, São Paulo, SP. 2007.

RICHARDSON, R.C.D. The wear of metallic materials by soil. Journal of Agricultural Engineering Research, v. 12, 1967, p.22-39.

RIGNEY, D. View point set on material aspects of wear: introduction. Scripta. Salvador, BA. (2009).

RODRIGUES, L. F., 2006, Comparative analysis of methodologies used to dispatch lorries in open-cast mines. Belo Horizonte, Minas Gerais, Brazil: Master's thesis;

STAFFORD, J. V.; TANNER, D. W.; The frictional characteristics of a Steel sliding on soil. Europe Journal of Soil Science, v.28, n.4, 1977, p.541-553.

MIX
Papier aus verantwortungsvollen Quellen
Paper from responsible sources
FSC® C105338

Printed by Books on Demand GmbH, Norderstedt / Germany